T0189878

MATERIALS SCIENCE AND ENGINEERING

Physical Process, Methods, and Models

Volume 1

MATERIALS SCIENCE AND ENGINEERING

Physical Process, Methods, and Models

Volume 1

Edited by

Abbas Hamrang, PhD

Gennady E. Zaikov, DSc, and A. K. Haghi, PhD

Reviewers and Advisory Board Members

Apple Academic Press Inc.
3333 Mistwell Crescent
Oakville, ON L6L 0A2
Canada

Apple Academic Press Inc.
9 Spinnaker Way
Waretown, NJ 08758
USA

First issued in paperback 2021

Exclusive worldwide distribution by CRC Press, a member of Taylor & Francis Group

ISBN 13: 978-1-77463-090-7 (pbk)
ISBN 13: 978-1-77188-000-8 (hbk)

Library of Congress Control Number: 2014937942

Library and Archives Canada Cataloguing in Publication

Materials science and engineering.

Includes bibliographical references and index.
Contents: Volume 1. Physical process, methods, and models/edited by Abbas Hamrang, PhD; Gennady E. Zaikov, DSc, and A.K. Haghi, PhD, Reviewers and Advisory Board Members -- Volume 2. Physiochemical concepts, properties, and treatments/edited by G.E. Zaikov, DSc, A.K. Haghi, PhD, and E. Kłodzińska PhD.
ISBN 978-1-77188-054-1 (set).--ISBN 978-1-77188-000-8 (v. 1 : bound).--
ISBN 978-1-77188-009-1 (v. 2 : bound)

1. Materials. I. Hamrang, Abbas, editor II. Zaikov, G. E. (Gennady Efremovich), 1935-, author, editor III. Haghi, A. K., author, editor IV. Kłodzińska, Ewa, editor V. Title: Physical process, methods, and models. VI. Title: Physiochemical concepts, properties, and treatments.

TA403.M38 2014 620.1'1 C2014-902356-1

ABOUT THE EDITOR

Abbas Hamrang, PhD

Abbas Hamrang, PhD, is a professor of polymer science and technology. He is currently a senior polymer consultant and editor and member of the academic boards of various international journals. His research interests include degradation studies of historical objects and archival materials, cellulose-based plastics, thermogravemetric analysis, accelerated ageing process and stabilization of polymers by chemical and non-chemical methods. His previous involvement in academic and industry sectors at the international level include Deputy Vice-Chancellor of Research and Development, Senior Lecturer, Manufacturing Consultant, and Science and Technology Advisor.

REVIEWERS AND ADVISORY BOARD MEMBERS

Gennady E. Zaikov, DSc

Gennady E. Zaikov, DSc, is Head of the Polymer Division at the N. M. Emanuel Institute of Biochemical Physics, Russian Academy of Sciences, Moscow, Russia, and Professor at Moscow State Academy of Fine Chemical Technology, Russia, as well as Professor at Kazan National Research Technological University, Kazan, Russia. He is also a prolific author, researcher, and lecturer. He has received several awards for his work, including the the Russian Federation Scholarship for Outstanding Scientists. He has been a member of many professional organizations and on the editorial boards of many international science journals.

A. K. Haghi, PhD

A. K. Haghi, PhD, holds a BSc in urban and environmental engineering from University of North Carolina (USA); a MSc in mechanical engineering from North Carolina A&T State University (USA); a DEA in applied mechanics, acoustics and materials from Université de Technologie de Compiègne (France); and a PhD in engineering sciences from Université de Franche-Comté (France). He is the author and editor of 65 books as well as 1000 published papers in various journals and conference proceedings. Dr. Haghi has received several grants, consulted for a number of major corporations, and is a frequent speaker to national and international audiences. Since 1983, he served as a professor at several universities. He is currently Editor-in-Chief of the *International Journal of Chemoinformatics and Chemical Engineering* and *Polymers Research Journal* and on the editorial boards of many international journals. He is a member of the Canadian Research and Development Center of Sciences and Cultures (CRDCSC), Montreal, Quebec, Canada.

CONTENTS

List of Contributors .... *xi*
List of Abbreviations .... *xiii*
List of Symbols .... *xv*
Preface .... *xvii*

1. **Advances in Electrospun Nanofibers Modeling: An Overview........ 1**
S. Rafiei, S. Maghsoodlou, B. Noroozi, and A. K. Haghi

2. **Affinity Separation of Enzymes Using Immobilized Metal Ions PGMA Grafted Cellophane Membranes: β-Galactosidase Enzyme Model........ 111**
M. S. Mohy Eldin, M. A. Abu-Saied, E.A. Soliman, and E.A. Hassan

3. **Satellite Imaging for Assessing the Annual Variation of Fish Catch in East and West Coast of India........ 137**
C. O. Mohan, B. Meenakumari, A. K. Mishra, D. Mitra, and T. K. Srinivasa Gopal

4. **Mechanisms of Catalysis with Binary and Triple Catalytic Systems.... 161**
L. I. Matienko, V. I. Binyukov, L. A. Mosolova, E. M. Mil, and G. E. Zaikov

5. **Synthesis of Synthetic Mineral-Based Alloys Liquation Phenomena of Differentiation........ 189**
A. M. Ignatova and M. N. Ignatov

6. **Restructuring of Synthetic Mineral Alloys Under Impact........ 199**
A. M. Ignatova and M. N. Ignatov

7. **Investigation of Efficiency of the Intumescent Fire and Heat Retardant Coatings Based on Perchlorovinyl Resin for Fiberglass Plastics........ 213**
V. F. Kablov, N. A. Keibal, S. N. Bondarenko, M. S. Lobanova, and A. N. Garashchenko

8. **Mechanical Performance Evaluation of Nanocomposite Modified Asphalts........ 225**
M. Arabani, and V. Shakeri

9. Microstructural Complexity of Natural and Synthetic Graphite Particles.. 281

Heinrich Badenhorst

Index... 281

LIST OF CONTRIBUTORS

M. A. Abu-Saied
Polymer Materials Research Department, Membranes' Applications Research Group, Advanced Technologies and New Materials Research Institute (ATNMRI), Scientific Research and Technological Applications City (SRTA-City), New Borg El-Arab City, Alexandria 21934, Egypt, E-mail: mouhamed-abdelrehem@yahoo.com

M. Arabani
School of Engineering, University of Guilan, Guilan, Rasht, IRAN, P.O. BOX: 3756, E-mail: arabani@guilan.ac.ir

Heinrich Badenhorst
SARChI Chair in Carbon Materials and Technology, Department of Chemical Engineering, University of Pretoria, Lynwood Road, Pretoria, Gauteng, 0002, South Africa / P.O. Box 66464, Highveld Ext. 7, Centurion, Gauteng, 0169, South Africa, Tel.: +27 12 420 4173; Fax: +27 12 420 2516; E-mail: heinrich.badenhorst@up.ac.za

V. I. Binyukov
N.M. Emanuel Institute of Biochemical Physics, Russian Academy of Sciences, ul. Kosygina 4, 119334 Moscow, Russian Federation. Tel.: (7-495) 939 71 40; Fax: (7-495) 137 41 01

S. N. Bondarenko
Volzhsky Polytechnical Institute (branch), Volgograd State Technical University, 42a Engelsa Street, Volzhsky, Volgograd Region, 404121, Russian Federation, E-mail: vtp@volpi.ru; www.volpi.ru

A. N. Garashchenko
Volzhsky Polytechnical Institute (branch), Volgograd State Technical University, 42a Engelsa Street, Volzhsky, Volgograd Region, 404121, Russian Federation, E-mail: vtp@volpi.ru; www.volpi.ru

T. K. Srinivasa Gopal
Central Institute of Fisheries Technology, Cochin

A. K. Haghi
University of Guilan, Rasht, Iran

E. A. Hassan
Department of Chemistry, Faculty of Science, Al-Azhar University, Cairo, Egypt

M. N. Ignatov
Mechanical Engineering Department, National Research Perm Polytechnic University, Perm, Russia

V. F. Kablov
Volzhsky Polytechnical Institute (branch), Volgograd State Technical University, 42a Engelsa Street, Volzhsky, Volgograd Region, 404121, Russian Federation, E-mail: vtp@volpi.ru; www.volpi.ru

N. A. Keibal
Volzhsky Polytechnical Institute (branch), Volgograd State Technical University, 42a Engelsa Street, Volzhsky, Volgograd Region, 404121, Russian Federation, E-mail: vtp@volpi.ru; www.volpi.ru

M. S. Lobanova
Volzhsky Polytechnical Institute (branch), Volgograd State Technical University, 42a Engelsa Street, Volzhsky, Volgograd Region, 404121, Russian Federation, E-mail: vtp@volpi.ru; www.volpi.ru

S. Maghsoodlou
University of Guilan, Rasht, Iran

L. I. Matienko
N.M. Emanuel Institute of Biochemical Physics, Russian Academy of Sciences, ul. Kosygina 4, 119334 Moscow, Russian Federation. Tel.: (7-495) 939 71 40; Fax: (7-495) 137 41 01; E-mail: matienko@sky.chph.ras.ru

B. Meenakumari
Director General (Fisheries), ICAR, New Delhi

E. M. Mil
N.M. Emanuel Institute of Biochemical Physics, Russian Academy of Sciences, ul. Kosygina 4, 119334 Moscow, Russian Federation. Tel.: (7-495) 939 71 40; Fax: (7-495) 137 41 01

A. K. Mishra
Marine and Atmospheric Sciences Department Indian Institute of Remote Sensing, Dehradun

D. Mitra
Marine and Atmospheric Sciences Department Indian Institute of Remote Sensing, Dehradun

C. O. Mohan
Veraval Research Centre of Central Institute of Fisheries Technology, Bhidia, Veraval, Gujarat, E-mail: comohan@gmail.com

M. S. Mohy Eldin
Polymer Materials Research Department, Membranes' Applications Research Group, Advanced Technologies and New Materials Research Institute (ATNMRI), Scientific Research and Technological Applications City (SRTA-City), New Borg El-Arab City, Alexandria 21934, Egypt

L. A. Mosolova
N.M. Emanuel Institute of Biochemical Physics, Russian Academy of Sciences, ul. Kosygina 4, 119334 Moscow, Russian Federation. Tel.: (7-495) 939 71 40; Fax: (7-495) 137 41 01

B. Noroozi
University of Guilan, Rasht, Iran

S. Rafiei
University of Guilan, Rasht, Iran

V. Shakeri
University of Guilan, Rasht, Iran

E. A. Soliman
Polymer Materials Research Department, Membranes' Applications Research Group, Advanced Technologies and New Materials Research Institute (ATNMRI), Scientific Research and Technological Applications City (SRTA-City), New Borg El-Arab City, Alexandria 21934, Egypt

G. E. Zaikov
N. M. Emanuel Institute of Biochemical Physics, Russian Academy of Sciences, ul. Kosygina 4, 119334 Moscow, Russian Federation. Tel.: (7-495) 939 71 40; Fax: (7-495) 137 41 01

LIST OF ABBREVIATIONS

18C6 18-crown-6
ACAC Acetylacetonate ion
AFM method Atomic-Force Microscopy method
ARD Acireductone ligand
ASA Active Surface Area
ATNMRI Advanced Technologies and New Materials Research Institute
BEM Boundary Element Method
BGK Bhatnagar-Gross-Krook
BSA Bovine Serum Albumin
BSQ Band Sequential Format
Bu Butyl radical
$CHCl_3$ Chloroform
CO Carbon monoxide
CTAB Cetyltrimethylammonium bromide ($Me_3(n\text{-}C_{16}H_{33})NBr$)
CZCS Coastal Zone Color Scanner
DMF Dimethylformamide
DSMC Direct Monte Carlo Simulation
EEZ Exclusive Economic Zone
Et Ethyl radical
GA Glycidyl Methacrylate
Hacac Acetylacetone
HDF Hierarchial Data Format
HMPA Hexamethylphosphorotriamide
IMAC Immobilized Metal-Ion Affinity Chromatography
KPS Potassium Persulphate
LEDs Light Emitting Diodes
MD Molecular Dynamics
Me Methyl radical
Mo Molybdenum
MODIS Moderate Resolution Imaging Spectroradiometer
MP N-methylpirrolidon-2

MPC	Methylphenylcarbinol
MSP	Methionine Salvage Pathway
MSt	Stearates of alkaline metals (M= Li, Na, K)
Ni(Fe)ARD	Ni(Fe)AcireductoneDioxygenase
NIR	Near Infrared
NNG	Nuclear grade Natural Graphite
NO	nitrogen monoxide
NSG	Nuclear grade Synthetic Graphite
OAc	Acetate ion
ODEs	Ordinary Differential Equations
PDEs	Partial Differential Equations
PEDA	Phosphorus-Boron-Nitrogen-Containing Oligomer
PEH	Phenyl ethyl hydro peroxide
PFZ	Potential Fishing Zones
QX	Quaternary ammonium salt
RH	Ethyl benzene
RLA	Repeated Load Axial
RV	Rotational Viscometer
SEM	Scanning Electron Microscopy
SST	Sea Surface Temperature
TEG	Thermal Expanded Graphite
TGA	Thermogravimetric Analyzer
UV-spectrum	Ultra Violet-spectrum
ZnO	Nano Zinc Oxide

LIST OF SYMBOLS

t_n^e axial viscous normal stress, N/m^2

t_t^e pressure, N/m^2

$\bar{l}$ mean segment length, m

χ aspect ratio

σ electric field

ϕ fiber orientation angle

ν jet velocity, m^3/s

y(j,l) length distribution

τ_{rr} dielectric constant of the jet, N/m^2

<QQ> suspension configuration tensor, N/m^2

A accumulation built up within the system

A(s), S current density, m^2

C consumption used in system volume

G generation produced in system volume

h distance from pendent drop to ground collector

I input entering through the system surface

J jet current

K Boltzmann's constant

m equivalent mass

O output leaving through system boundary

P' polarization

P spinning distance

Q flow rate, m^3/s

R' jet axial position

R radius of jet, m

R_0 initial radius

U_e normal electric force, V

v kinematic viscosity

V_0 applied voltage

W_e tangential electric force, M

β Galactosidase and Bovine Serum Albumin

γ surface tension of the fluid

ε' Lagrangian axial strain

ε□ slope of the jet surface, J

ε surface tension

ρ density of the fluid

σ_V viscoelastic stress

τ relaxation time of the polymer

PREFACE

This volume highlights the latest developments and trends in materials science and engineering. It presents the developments of advanced materials and respective tools to characterize and predict the material properties and behavior. This book has an important role in advancing materials science and engineering in macro and nanoscale. Its aim is to provide original, theoretical, and important experimental results that use non-routine methodologies often unfamiliar to the usual readers. It also includes chapters on novel applications of more familiar experimental techniques and analyses of composite problems that indicate the need for new experimental approaches.

The book is for professors and instructors of specific teaching courses, students and postgraduate students focusing on adhesive interaction improvement, and industry professionals working in materials science.

In chapter 1, a new approach in nanostructured materials is a computational-based material is developed. This is based on multiscale material and process modeling spanning on a large spectrum of time as well as on length scales. The cost of designing and producing novel multifunctional nanomaterials can be high and the risk of investment can be significant. Computational nanomaterials research that relies on multiscale modeling has the potential to significantly reduce development costs of new nanostructured materials for demanding applications by bringing physical and microstructural information into the realm of the design engineer.

Chapter 2 is focused on metals immobilization, and selected membranes with highest sulphonation degree were immobilized.

The study presented in chapter 3 was undertaken with the objective of evaluating the correlation of the chlorophyll a and sea surface temperature derived from the satellite MODIS.

Mechanisms of catalysis with binary and triple catalytic systems is investigated in chapter 4.

Synthesis of synthetic mineral-based alloys liquation phenomena of differentiation is reviewed in chapter 5.

The aim of chapter 6 was to study structural changes in siminals, specifically raw hornblendite materials, under shock impact.

The purpose of chapter 7 was obtaining fire-retardant coatings based on perchlorovinyl resin with improved adhesive properties to protect fiberglass plastics. The chapter presents the results of studies on the influence of a modifier based on the phosphorus-boron-nitrogen-containing oligomer (PEDA) and filler, which is thermal-expanded graphite on physical, mechanical and fire retardant properties of the coatings.

The goal of chapter 8 is to evaluate the influence of nano ZnO on the engineering properties of bitumen and asphalt concrete mixtures. For this purpose, the authors performed penetration grade, softening point, ductility, and rotational viscometer (RV) tests on modified bitumen by four different content of nano ZnO and repeated load axial (RLA) test on asphalt concrete mixtures by three different content of nano ZnO. With the experimental results and the numerical analysis with Matlab Software, two experimental models were proposed for prediction of the creep behavior of both conventional and modified asphalt mixtures with optimum nano ZnO for different conditions depending on temperature and stress.

Microstructural complexity of natural and synthetic graphite particles is reviewed in chapter 9.

—Abbas Hamrang, PhD

CHAPTER 1

ADVANCES IN ELECTROSPUN NANOFIBERS MODELING: AN OVERVIEW

S. RAFIEI, S. MAGHSOODLOU, B. NOROOZI, and A. K. HAGHI

CONTENTS

1.1 An Introduction to Nanotechnology 2
1.2 Nanostructured Materials 5
1.3 Nanofiber Technology 14
1.4 Introduction to Theoretical Study of Electrospinning Process 25
1.5 Study of Electrospinning Jet Path 27
1.6 Electrospinning Drawbacks 31
1.7 Modelling of The Electrospinning Process 33
1.8 Electrospinning Simulation 75
1.9 Electrospinning Simulation Example 75
1.10 Applied Numerical Methods gor Electrospinning 80
1.11 Conclusion 96
Keywords 97
References 98

1.1 AN INTRODUCTION TO NANOTECHNOLOGY

The nanostructure materials productions are most challenging and innovative processes, introducing, in the manufacturing, a new approaches such as self-assembly and self-replication. The fast growing of nanotechnology with modern computational/experimental methods gives the possibility to design multifunctional materials and products in human surroundings. Some of them are smart clothing, portable fuel cells, and medical devices are some of them. Research in nanotechnology began with applications outside of everyday life and is based on discoveries in physics and chemistry. The reason for that is needed to understand the physical and chemical properties of molecules and nanostructures in order to control them.

A new approach in nanostructured materials is a computational-based material development. It is based on multiscale material and process modeling spanning, on a large spectrum of time as well as on length scales. The cost of designing and producing novel multifunctional nanomaterials can be high and the risk of investment to be significant. Computational nanomaterials research that relies on multiscale modeling has the potential to significantly reduce development costs of new nanostructured materials for demanding applications by bringing physical and microstructural information into the realm of the design engineer.

One of the most significant types of these one-dimensional nanomaterials is nanofibers, which can be produced widely through electrospinning procedure. A drawback of this method, however, is the unstable behavior of the liquid jet, which causes the fibers to be collected randomly. So a critical concern in this process is to achieve desirable control. Studying the dynamics of electrospinning jet would be easier and faster if it can be modeled and simulated, rather than doing experiments. This chapter focuses on modeling and then simulating of electrospinning process in various views. In order to study the applicability of the electrospinning modeling equations, which discussed in detail in earlier parts of this approach, an existing mathematical model in which the jet was considered as a mechanical system was interconnected with viscoelastic elements and used to build a numeric method. The simulation features the possibility of predicting essential parameters of electrospinning process and the results have good agreement with other numeric studies of electrospinning, which modeled this process based on axial direction.

Understanding the nanoworld makes up one of the frontiers of modern science. One reason for this is that technology based on nanostructures promises to be hugely important economically [1–3]. Nanotechnology literally means any technology on a nanoscale that has applications in the real world. It includes the production and application of physical, chemical, and biological systems at scales ranging from individual atoms or molecules to submicron dimensions, as well as the integration of the resulting nanostructures into larger systems. Nanotechnology is likely to have a profound impact on our economy and society in the early twenty-first century, comparable to that of semiconductor technology, information technology, or cellular and molecular biology. Science and technology research in nanotechnology promises breakthroughs in areas such as materials and manufacturing [4], nanoelectronics [5], medicine and healthcare [6], energy [7], biotechnology [8], information technology [9], and national security [10]. It is widely felt that nanotechnology will be the next Industrial Revolution [9].

As far as "nanostructures" are concerned, one can view this as objects or structures whereby at least one of its dimensions is within nano-scale. A "nanoparticle" can be considered as a zero dimensional nano-element, which is the simplest form of nanostructure. It follows that a "nanotube" or a nanorod" is a one-dimensional nano-element from which slightly more complex nanostructure can be constructed of [11–12].

Following this fact, a "nanoplatelet" or a "nanodisk" is a two-dimensional element, along with its one-dimensional counterpart, which is useful in the construction of nanodevices. The difference between a nanostructure and a nanodevice can be viewed upon as the analogy between a building and a machine (whether mechanical, electrical or both) [1]. It is important to know that as far as nanoscale is concerned; these nano-elements should not consider only as an element that form a structure while they can be used as a significant part of a device. For example, the use of carbon nanotube as the tip of an Atomic Force Microscope (AFM) would have it classified as a nanostructure. The same nanotube, however, can be used as a singlemolecule circuit, or as part of a miniaturized electronic component, thereby appearing as a nanodevice. Therefore the function, along with the structure, is essential in classifying which nanotechnology subarea it belongs to. This classification will be discussed in detail in further sections [11, 13].

As long as nanostructures clearly define the solids' overall dimensions, the same cannot be said so for nanomaterials. In some instances a nanomaterial refers to a nano-sized material while in other instances a nanomaterial is a bulk material with nano-scaled structures. Nanocrystals are other groups of nanostructured materials. It is understood that a crystal is highly structured and that the repetitive unit is indeed small enough. Hence a nanocrystal refers to the size of the entire crystal itself being nano-sized, but not of the repetitive unit [14].

Nanomagnetics are the other type of nanostructured materials, which are known as highly miniaturized magnetic data storage materials with very high memory. This can be attained by taking advantage of the electron spin for memory storage, hence the term "spin-electronics," which has since been more popularly and more conveniently known as "spintronics" [1, 9, 15]. In nanobioengineering, the novel properties of nano-scale are taken advantage of for bioengineering applications. The many naturally occurring nanofibrous and nanoporous structure in the human body further adds to the impetus for research and development in this subarea. Closely related to this is molecular functionalization whereby the surface of an object is modified by attaching certain molecules to enable desired functions to be carried out such as for sensing or filtering chemicals based on molecular affinity [16–17].

With the rapid growth of nanotechnology, nanomechanics are no longer the narrow field, which it used to be [13]. This field can be broadly categorized into the molecular mechanics and the continuum mechanics approaches which view objects as consisting of discrete many-body system and continuous media, respectively. As long as the former inherently includes the size effect, it is a requirement for the latter to factor in the influence of increasing surface-to-volume ratio, molecular reorientation and other novelties as the size shrinks. As with many other fields, nanotechnology includes nanoprocessing novel materials processing techniques by which nano-scale structures and devices are designed and constructed [18, 19].

Depending on the final size and shape, a nanostructure or nanodevice can be created from the top-down or the bottom-up approach. The former refers to the act of removal or cutting down a bulk to the desired size, while the latter takes on the philosophy of using the fundamental building blocks—such as, atoms and molecules, to build up nanostructures in the same manner. It is obvious that the top-down and the bottom-up nanoprocessing methodologies are suitable for the larger and two smaller ends, respectively, in the spectrum of nano-scale construction. The effort of nanopatterning—or patterning at the nanoscale would therefore fall into nanoprocessing [1, 12, 18].

1.2 NANOSTRUCTURED MATERIALS

Strictly speaking, a nanostructure is any structure with one or more dimensions measuring in the nanometer (10^{-9} m) range. Various definitions refine this further, stating that a nanostructure should have a characteristic dimension lying between 1 nm and 100 nm, putting nanostructures as intermediate in size between a molecule and a bacterium. Nanostructures are typically probed either optically (spectroscopy, photoluminescence, etc.) or in transport experiments. This field of investigation is often given the name mesoscopic transport, and the following considerations give an idea of the significance of this term [1, 2, 12, 20, 21].

What makes nanostructured materials very interesting and award them with their unique properties is that their size is smaller than critical lengths that characterize many physical phenomena. Generally, physical properties of materials can be characterized by some critical length, for example, a thermal diffusion length, or a scattering length. The electrical conductivity of a metal is strongly determined by the distance that the electrons travel between collisions with the vibrating atoms or impurities of the solid. This distance is called the mean free path or the scattering length. If the sizes of the particles are less than these characteristic lengths, it is possible that new physics or chemistry may occur [1, 9, 17].

Several computational techniques have been employed to simulate and model nanomaterials. Since the relaxation times can vary anywhere from picoseconds to hours, it becomes necessary to employ Langevin dynamics besides molecular dynamics in the calculations. Simulation of nanodevices through the optimization of various components and functions provides challenging and useful task [20, 22]. There are many examples where simulation and modeling have yielded impressive results, such as nanoscale lubrication

[23]. Simulation of the molecular dynamics of DNA has been successful to some extent [24]. Quantum dots and nanotubes have been modeled satisfactorily [25, 26]. First principles calculations of nanomaterials can be problematic if the clusters are too large to be treated by Hartree–Fock methods and too small for density functional theory [1]. In the next section various classifications of these kinds of materials are considered in detail.

1.2.1 CLASSIFICATION OF NANOSTRUCTURED MATERIALS

Nanostructure materials as a subject of nanotechnology are low dimensional materials comprising of building units of a submicron or nanoscale size at least in one direction and exhibiting size effects. The first classification idea of NSMs was given by Gleiter in 1995 [3]. A modified classification scheme for these materials, in which 0D, 1D, 2D and 3D dimensions are included suggested in later researches [21]. These classifications are as follows:

1.2.1.1 0D NANOPARTICLES

A major feature that distinguishes various types of nanostructures is their dimensionality. In the past 10 years, significant progress has been made in the field of zero-dimension nanostructure materials. A rich variety of physical and chemical methods have been developed for fabricating these materials with well-controlled dimensions [3, 18]. Recently, 0D nanostructured materials such as uniform particles arrays (quantum dots), heterogeneous particles arrays, core–shell quantum dots, onions, hollow spheres and nanolenses have been synthesized by several research groups [21]. They have been extensively studied in light emitting diodes (LEDs), solar cells, single-electron transistors, and lasers.

1.2.1.2 1D NANOPARTICLES

In the last decade, 1D nanostructured materials have focused an increasing interest due to their importance in research and developments and have a wide range of potential applications [27]. It is generally accepted that these materials are ideal systems for exploring a large number of novel phenomena at the nanoscale and investigating the size and dimensionality dependence of functional properties. They are also expected to play an important role as both interconnects and the key units in fabricating electronic, optoelectronic, and EEDs with nanoscale dimensions. The most important types of this group are nanowires, nanorods, nanotubes, nanobelts, nanoribbons, hierarchical nanostructures and nanofibers [1, 18, 28].

1.2.1.3 2D NANOPARTICLES

2D nanostructures have two dimensions outside of the nanometric size range. In recent years, synthesis of 2D nanomaterial has become a focal area in materials research, owing to their many low dimensional characteristics different from the bulk properties. Considerable research attention has been focused over the past few years on the development of them. 2D nanostructured materials with certain geometries exhibit unique shape-dependent characteristics and subsequent utilization as building blocks for the key components of nanodevices [21]. In addition, these materials are particularly interesting not only for basic understanding of the mechanism of nanostructure growth, but also for investigation and developing novel applications in sensors, photocatalysts, nanocontainers, nanoreactors, and templates for 2D structures of other materials. Some of the three-dimension nanoparticles are junctions (continuous islands), branched structures, nanoprisms, nanoplates, nanosheets, nanowalls, and nanodisks [1].

1.2.1.4 3D NANOPARTICLES

Owing to the large specific surface area and other superior properties over their bulk counterparts' arising from quantum size effect, they have attracted considerable research interest and many of them have been synthesized in the past 10 years [1, 12]. It is well known that the behaviors of NSMs strongly depend on the sizes, shapes, dimensionality and morphologies, which are thus the key factors to their ultimate performance and applications. Therefore, it is of great interest to synthesize 3D NSMs with a controlled structure and morphology. In addition, 3D nanostructures are an important material due to its wide range of applications in the area of catalysis, magnetic material and electrode material for batteries [2]. Moreover, the 3D NSMs have recently attracted intensive research interests because the nanostructures have higher surface area and supply enough absorption sites for all involved molecules in a small space [58]. On the other hand, such materials with porosity in three dimensions could lead to a better transport of the molecules. Nanoballs (dendritic structures), nanocoils, nanocones, nanopillers and nanoflowers are in this group [1, 2, 18, 29].

1.2.2 SYNTHESIS METHODS OF NANOMATERIALS

The synthesis of nanomaterials includes control of size, shape, and structure. Assembling the nanostructures into ordered arrays often becomes necessary for rendering them functional and operational. In the last decade, nanoparticles (powders) of ceramic materials have been produced in large scales by employing both physical and chemical methods. There has been considerable progress in the preparation of nanocrystals of metals, semiconductors, and magnetic materials by employing colloid chemical methods [18, 30].

The construction of ordered arrays of nanostructures by employing techniques of organic self-assembly provides alternative strategies for nanodevices. Two and three-dimensional arrays of nanocrystals of semiconductors, metals, and magnetic materials have been assembled by using suitable organic reagents [1, 31]. Strain directed assembly of nanoparticle arrays (e.g., of semiconductors) provides the means to introduce functionality into the substrate that is coupled to that on the surface [32].

Preparation of nanoparticles is an important branch of the materials science and engineering. The study of nanoparticles relates various scientific fields, for example, chemistry, physics, optics, electronics, magnetism and mechanism of materials. Some nanoparticles have already reached practical stage. In order to meet the nanotechnology and nano-materials development in the next century, it is necessary to review the preparation techniques of nanoparticles.

All particle synthesis techniques fall into one of the three categories: vapor-phase, solution precipitation, and solid-state processes. Although vapor-phase processes have been common during the early days of nanoparticles development, the last of the three processes mentioned above is the most widely used in the industry for production of micron-sized particles, predominantly due to cost considerations [18, 33].

Methods for preparation of nanoparticles can be divided into physical and chemical methods based on whether there exist chemical reactions [33]. On the other hand, in general, these methods can be classified into the gas phase, liquid phase and solid phase methods based on the state of the reaction system. The gas phase method includes gas-phase evaporation method (resistance heating, high frequency induction heating, plasma heating, electron beam heating, laser heating, electric heating evaporation method, vacuum deposition on the surface of flowing oil and exploding wire method), chemical vapor reaction (heating heat pipe gas reaction, laser induced chemical vapor reaction, plasma enhanced chemical vapor reaction), chemical vapor condensation and sputtering method. Liquid phase method for synthesizing nanoparticles mainly includes precipitation, hydrolysis, spray, solvent thermal method (high temperature and high pressure), solvent evaporation pyrolysis, oxidation-reduction (room pressure), emulsion, radiation chemical synthesis and sol-gel processing. The solid phase method includes thermal decomposition, solid-state reaction, spark discharge, stripping, and milling method [30, 33].

In other classification, there are two general approaches to the synthesis of nanomaterials and the fabrication of nanostructures, bottom–up and top–down approach. The first one includes the miniaturization of material components (up to atomic level) with further self-assembly process leading to the formation assembly of nanostructures. During self-assembly the physical forces operating at nanoscale are used to combine basic units into larger stable structures. Typical examples are quantum dot formation during epitaxial growth and formation of nanoparticles from colloidal dispersion. The latter uses larger (macroscopic) initial structures, which can be externally controlled in the processing of nanostructures. Typical examples are etching through the mask, ball milling, and application of severe plastic deformation [3, 13].

Some of the most common methods are described in following:

1.2.2.1 PLASMA BASED METHODS

Metallic, semiconductive and ceramic nanomaterials are widely synthesized by hot and cold plasma methods. A plasma is sometimes referred to as being "hot" if it is nearly fully ionized, or "cold" if only a small fraction, (for instance 1%), of the gas molecules are ionized, but other definitions of the terms "hot plasma" and "cold plasma" are common. Even in cold plasma, the electron temperature is still typically several thousand centigrades. Generally, the related equipment consists of an arc melting chamber and a collecting system. The thin films of alloys were prepared from highly pure metals by arc melting in an inert gas atmosphere. Each arc-melted ingot was flipped over and remelted three times. Then, the thin films of alloy were produced by arc melting a piece of bulk materials in a mixing gas atmosphere at a low pressure. Before the ultrafine particles were taken out from the arc-melting chamber, they were passivized with a mixture of inert gas and air to prevent the particles from burning up [34–35].

Cold plasma method is used for producing nanowires in large scale and bulk quantity. The general equipment of this method consists of a conventional horizontal quartz tube furnace and an inductively coupled coil driven by a 13.56 MHz radio-frequency (RF or radio-frequency) power supply. This method often is called as an RF plasma method. During RF plasma method, the starting metal is contained in a pestle in an evacuated chamber. The metal is heated above its evaporation point using high voltage RF coils wrapped around the evacuated system in the vicinity of the pestle. Helium gas is then allowed to enter the system, forming high temperature plasma in the region of the coils. The metal vapor nucleates on the He gas atoms and diffuses up to a colder collector rod where nanoparticles are formed. The particles are generally passivized by the introduction of some gas such as oxygen. In the case of aluminum nanoparticles the oxygen forms a layer of aluminum oxide about the particle [1, 36].

1.2.2.2 CHEMICAL METHODS

Chemical methods have played a major role in developing materials imparting technologically important properties through structuring the materials on the nanoscale. However, the primary advantage of chemical processing is its versatility in designing and synthesizing new materials that can be refined into the final end products. The secondary most advantage that the chemical processes offer over physical methods is a good chemical homogeneity, as a chemical method offers mixing at the molecular level. On the other hand, chemical methods frequently involve toxic reagents and solvents for the synthesis of nanostructured materials. Another disadvantage of the chemical methods is the unavoidable introduction of byproducts, which require subsequent purification steps after the synthesis, in other words, this process is time consuming. In spite of these facts, probably the most useful methods of synthesis in terms of their potential to be scaled up are chemical methods [33, 37]. There are a number of different chemical methods that can be used to make nanoparticles of metals, and we will give some examples. Several types of reducing agents can be used to produce nanoparticles such as $NaBEt_3H$, $LiBEt_3H$, and $NaBH_4$ where Et denotes the ethyl ($-C_2H_5$) radical. For example, nanoparticles of molybdenum (Mo) can be reduced in toluene solution with $NaBEt_3H$ at room temperature, providing a high yield of Mo nanoparticles having dimensions of 1–5 nm [30].

1.2.2.3 THERMOLYSIS AND PYROLYSIS

Nanoparticles can be made by decomposing solids at high temperature having metal cations, and molecular anions or metal organic compounds. The process is called thermolysis. For example, small lithium particles can be made by decomposing lithium oxide, LiN_3. The material is placed in an evacuated quartz tube and heated to 400°C in the apparatus. At about 370°C the LiN_3 decomposes, releasing N_2 gas, which is observed by an increase in the pressure on the vacuum gauge. In a few minutes the pressure drops back to its original low value, indicating that all the N_2 has been removed. The remaining lithium atoms coalesce to form small colloidal metal particles. Particles less than 5 nm can be made by this method. Passivation can be achieved by introducing an appropriate gas [1].

Pyrolysis is commonly a solution process in which nanoparticles are directly deposited by spraying a solution on a heated substrate surface, where the constituent reacts to form a chemical compound. The chemical reactants are selected such that the products other than the desired compound are volatile at the temperature of deposition. This method represents a very simple and relatively cost-effective processing method (particularly in regard to equipment costs) as compared to many other film deposition techniques [30].

The other pyrolysis-based method that can be applied in nanostructures production is a laser pyrolysis technique, which requires the presence in the reaction medium of a molecule absorbing the CO_2 laser radiation [38, 39]. In most cases, the atoms of a molecule are rapidly heated via vibrational excitation and are dissociated. But in some cases, a sensitizer gas such as SF_6 can be directly used. The heated gas molecules transfer their energy to the reaction medium by collisions leading to dissociation of the reactive medium without, in the ideal case, dissociation of this molecule. Rapid thermalization occurs after dissociation of the reactants due to transfer collision. Nucleation and growth of NSMs can take place in the as-formed supersaturated vapor. The nucleation and growth period is very short time (0.1–10 ms). Therefore, the growth is rapidly stopped as soon as the particles leave the reaction zone. The flame-excited luminescence is observed in the reaction region where the laser beam intersects the reactant gas stream. Since there is no interaction with any walls, the purity of the desired products is limited by the purity of the reactants. However, because of the very limited size of the reaction zone with a faster cooling rate, the powders obtained in this wellness reactor present a low degree of agglomeration. The particle size is small (~5–50 nm range) with a narrow size distribution. Moreover, the average size can be manipulated by optimizing the flow rate, and, therefore, the residence time in the reaction zone [39, 40].

1.2.2.4 LASER BASED METHODS

The most important laser based techniques in the synthesis of nanoparticles is pulsed laser ablation. As a physical gas-phase method for preparing nanosized particles, pulsed laser ablation has become a popular method to prepare high-purity and ultra-fine nanomaterials of any composition

[41, 42]. In this method, the material is evaporated using pulsed laser in a chamber filled with a known amount of a reagent gas and by controlling condensation of nanoparticles onto the support. It's possible to prepare nanoparticles of mixed molecular composition such as mixed oxides/nitrides and carbides/nitrides or mixtures of oxides of various metals by this method. This method is capable of a high rate of production of 2–3 g/min [40].

Laser chemical vapor deposition method is the next laser-based technique in which photoinduced processes are used to initiate the chemical reaction. During this method, three kinds of activation should be considered. First, if the thermalization of the laser energy is faster than the chemical reaction, pyrolytic and/or photothermal activation is responsible for the activation. Secondly, if the first chemical reaction step is faster than the thermalization, photolytical (nonthermal) processes are responsible for the excitation energy. Thirdly, combinations of the different types of activation are often encountered. During this technique a high intensity laser beam is incident on a metal rod, causing evaporation of atoms from the surface of the metal. The atoms are then swept away by a burst of helium and passed through an orifice into a vacuum where the expansion of the gas causes cooling and formation of clusters of the metal atoms. These clusters are then ionized by UV radiation and passed into a mass spectrometer that measures their mass: charge ratio [1, 41–43].

Laser-produced nanoparticles have found many applications in medicine, bio-photonics, in the development of sensors, new materials and solar cells. Laser interactions provide a possibility of chemical clean synthesis, which is difficult to achieve under more conventional NP production conditions [42]. Moreover, a careful optimization of the experimental conditions can allow a control over size distributions of the produced nanoclusters. Therefore, many studies were focused on the investigation the laser nanofabrication. In particular, many experiments were performed to demonstrate nanoparticles formation in vacuum, in the presence of a gas or a liquid. Nevertheless, it is still difficult to control the properties of the produced particles. It is believed that numerical calculations can help to explain experimental results and to better understand the mechanisms involved [43].

Despite rapid development in laser physics, one of the fundamental questions still concern the definition of proper ablation mechanisms and the processes leading to the nano particles formation. Apparently, the progress in laser systems implies several important changes in these mechanisms, which depend on both laser parameters and material properties. Among the more studied ablation mechanisms there are thermal, photochemical and photomechanical ablation processes. Frequently, however, the mechanisms are mixed, so that the existing analytical equations are hardly applicable. Therefore, numerical simulation is needed to better understand and to optimize the ablation process [44].

So far, thermal models are commonly used to describe nanosecond (and longer) laser ablation. In these models, the laser-irradiated material experiences heating, melting, boiling and evaporation. In this way, three numerical approaches were used [29, 45]:

– **Atomistic approach** based on such methods as Molecular Dynamics (MD) and Direct Monte Carlo Simulation (DSMC). Typical calculation results provide detailed information about atomic positions, velocities, kinetic and potential energy;

– **Macroscopic approach** based hydrodynamic models. These models allow the investigations of the role of the laser-induced pressure gradient, which is particularly important for ultra-short laser pulses. The models are based on a one fluid two-temperature approximation and a set of additional models (equation of state) that determines thermal properties of the target;

– **Multi-scale approach** based on the combination of two approaches cited above was developed by several groups and was shown to be particularly suitable for laser applications.

1.3 NANOFIBER TECHNOLOGY

Nano fiber consists of two terms “Nano” and “fiber,” as the latter term is looking more familiar. Anatomists observed fibers as any of the filament constituting the extracellular matrix of connective tissue, or any elongated cells or thread like structures, muscle fiber or nerve fiber. According to textile industry fiber is a natural or synthetic filament, such as cotton or nylon, capable of being spun into simply as materials made of such filaments. Physiologists and biochemists use the term fiber for indigestible plant matter consisting of polysaccharides such as cellulose, that when eaten stimulates intestinal peristalsis. Historically, the term fiber or "Fiber" in British English comes from Latin "fibra." Fiber is a slender, elongated thread like structure. Nano is originated from Greek word "nanos" or "nannos" refer to "little old man" or "dwarf." The prefixes "nannos" or "nano" as nannoplanktons or nanoplanktons used for very small planktons measuring 2 to 20 micrometers. In modern "nano" is used for describing various physical quantities within the scale of a billionth as nanometer (length), nanosecond (time), nanogram (weight) and nanofarad (charge) [1, 4, 9, 46]. As it was mentioned before, nanotechnology refers to the science and engineering concerning materials, structures and devices which has at least one dimension is 100 nm or less. This term also refers for a fabrication technology, where molecules, specification and individual atoms, which have at least one dimension in nanometers or less is used to design or built objects. Nano fiber, as the name suggests the fiber having a diameter range in nanometer. Fibrous structure having at least one dimension in nanometer or less is defined as Nano fiber according to National Science Foundation (NSC). The term Nano describes the diameter of the fibrous shape at anything below one micron or 1000 nm [4, 18].

Nanofiber technology is a branch of nanotechnology whose primary objective is to create materials in the form of nanoscale fibers in order to achieve superior functions [1, 2, 4]. The unique combination of high specific surface area, flexibility, and superior directional strength makes such fibers a preferred material form for many applications ranging from clothing to reinforcements for aerospace structures. Indeed, while the primary classification of nanofibers is that of nanostructure or nanomaterial, other aspects of nanofibers such as its characteristics, modeling, application and processing would enable nanofibers to penetrate into many subfields of nanotechnology [4, 46, 47].

It is obvious that nanofibers would geometrically fall into the category of one dimensional nano-scale elements that includes nanotubes and nanorods. However, the flexible nature of nanofibers would align it along with other highly flexible nano-elements such as globular molecules (assumed as zero dimensional soft matter), as well as solid and liquid films of nanothickness (two dimensional). A nanofiber is a nanomaterial in view of its diameter, and can be considered a nanostructured material if filled with nanoparticles to form composite nanofibers [1, 48].

The study of the nanofiber mechanical properties as a result of manufacturing techniques, constituent materials, processing parameters and other factors would fall into the category of nanomechanics. Indeed, while the primary classification of nanofibers is that of nanostructure or nanomaterial, other aspects of nanofibers such as its characteristics, modeling, application and processing would enable nanofibers to penetrate into many subfields of nanotechnology [1, 18].

Although, the effect of fiber diameter on the performance and processibility of fibrous structures has long been recognized, the practical generation of fibers at the nanometer scale was not realized until the rediscovery and popularization of the electrospinning technology by Professor Darrell Reneker almost a decade ago [49, 50]. The ability to create nanoscale fibers from a broad range of polymeric materials in a relatively simple manner using the electrospinning process, coupled with the rapid growth of nanotechnology in recent years have greatly accelerated the growth of nanofiber technology. Although, there are several alternative methods for generating fibers in a nanometer scale, none matches the popularity of the electrospinning technology due largely to the simplicity of the electrospinning process [18]. These methods will be discussed in following sections.

1.3.1 VARIOUS NANOFIBERS PRODUCTION METHODS

As it was discussed in detail, nanofiber is defined as the fiber having at least one dimension in nanometer range which can be used for a wide range of medical applications for drug delivery systems, scaffold formation, wound healing and widely used in tissue engineering, skeletal tissue, bone tissue, cartilage tissue, ligament tissue, blood vessel tissue, neural tissue, etc. It is also used in dental and orthopedic implants [4, 51, 52]. Nano fiber can be formed using different techniques including: drawing, template synthesis, phases separation, self-assembly and electrospinning.

1.3.1.1 DRAWING

In 1998, nanofibers were fabricated with citrate molecules through the process of drawing for the first time [53]. During drawing process, the fibers are fabricated by contacting a previously deposited polymer solution droplet with a sharp tip and drawing it as a liquid fiber, which is then solidified by rapid evaporation of the solvent due to the high surface area. The drawn fiber can be connected to another previously deposited polymer solution droplet thus forming a suspended fiber. Here, the predeposition of droplets significantly limits the ability to extend this technique, especially in three-dimensional configurations and hard to access spatial geometries. Furthermore, there is a specific time in which the fibers can be pulled. The viscosity of the droplet continuously increases with time due to solvent evaporation from the deposited droplet. The continual shrinkage in the volume of the polymer solution droplet affects the diameter of the fiber drawn and limits the continuous drawing of fibers [54].

To overcome the above-mentioned limitation is appropriate to use hollow glass micropipettes with a continuous polymer dosage. It provides greater flexibility in drawing continuous fibers in any configuration. Moreover, this method offers increased flexibility in the control of key parameters of drawing such as waiting time before drawing (due to the required viscosity of the polymer edge drops), the drawing speed or viscosity, thus enabling repeatability and control on the dimensions of the fabricated fibers. Thus, drawing process requires a viscoelastic material that can undergo strong deformations while being cohesive enough to support the stresses developed during pulling [54–55].

1.3.1.2 TEMPLATE SYNTHESIS

Template synthesis implies the use of a template or mold to obtain a desired material or structure. Hence the casting method and DNA replication can be considered as template-based synthesis. In the case of nanofiber creation by [56], the template refers to a metal oxide membrane with through-thickness pores of nano-scale diameter. Under the application of water pressure on one side and restrain from the porous membrane causes extrusion of the polymer which, upon coming into contact with a solidifying solution, gives rise to nanofibers whose diameters are determined by the pores [1, 57].

This method is an effective route to synthesize nanofibrils and nanotubes of various polymers. The advantage of the template synthesis method is that the length and diameter of the polymer fibers and tubes can be controlled by the selected porous membrane, which results in more regular nanostructures. General feature of the conventional template method is that the membrane should be soluble so that it can be removed after synthesis in order to obtain single fibers or tubes. This restricts practical application of this method and gives rise to a need for other techniques [1, 56, 57].

1.3.1.3 PHASE SEPARATION METHOD

This method consists of five basic steps: polymer dissolution, gelation, solvent extraction, freezing and freeze-drying. In this process, it is observed that gelatin is the most difficult step to control the porous morphology of nano fiber. Duration of gelation varied with polymer concentration and gelation temperature. At low gelation temperature, nano-scale fiber network is formed, whereas, high gelation temperature led to the formation of platelet-like structure. Uniform nano fiber can be produced as the cooling rate is increased, polymer concentration affects the properties of nano fiber, as polymer concentration is increased porosity of fiber decreased and mechanical properties of fiber are increased [1, 58].

1.3.1.4 SELF-ASSEMBLY

Self–assembly refers to the build-up of nano scale fibers using smaller molecules. In this technique, a small molecule is arranged in a concentric manner so that they can form bonds among the concentrically arranged small molecules, which upon extension in the plane's normal gives the longitudinal axis of a nano fiber. The main mechanism for a generic self-assembly is the inter-molecular forces that bring the smaller unit together.

A hydrophobic core of alkyl residues and a hydrophilic exterior lined by peptide residues was found in obtained fiber. It is observed that the nano fibers produced with this technique have diameter range 5–8 mm approximately and several microns in length [1, 59].

Although there are a number of techniques used for the synthesis of nanofiber but electrospinning represents an attractive technique to fabricate polymeric biomaterial into nanofibers. Electrospinning is one of the most commonly used methods for the production of nanofiber. It has a wide advantage over the previously available fiber formation techniques because here electrostatic force is used instead of conventionally used mechanical force for the formation of fibers. This method will be debated comprehensively in the following sub-sections.

1.3.1.5 ELECTROSPINNING OF NANOFIBERS

Electrospinning is a straightforward and cost-effective method to produce novel fibers with diameters in the range of from less than 3 nm to over 1 mm, which overlaps contemporary textile fiber technology. During this process, an electrostatic force is applied to a polymeric solution to produce nanofiber [60, 61] with diameter ranging from 50 nm to 1000 nm or greater [49, 62, 63]; Due to surface tension the solution is held at the tip of syringe. Polymer solution is charged due to applied electric force. In the polymer solution, a force is induced due to mutual charge repulsion that is directly opposite to the surface tension of the polymer solution. Further increases in the electrical potential led to the elongation of the hemispherical surface of the solution at the tip of the syringe to form a conical shape known as "Taylor cone" [50, 64]. The electric potential is increased to overcome the surface tension forces to cause the formation of a jet, ejects from the tip of the Taylor cone. Due to elongation and solvent evaporation, charged jet instable and gradually thins in air primarily [62, 65–67]. The charged jet forms randomly oriented nanofibers that can be collected on a stationary or rotating grounded metallic collector [50]. Electrospinning provides a good method and a practical way to produce polymer fibers with diameters ranging from 40–2000 nm [49–50].

1.3.1.5.1 THE HISTORY OF ELECTROSPINNING METHODOLOGY

William Gilbert discovered the first record of the electrostatic attraction of a liquid in 1600 [68]. The first electrospinning patent was submitted by John Francis Cooley in 1900 [69]. After that in 1914 John Zeleny studied on the behavior of fluid droplets at the end of metal capillaries, which caused the beginning of the mathematical model the behavior of fluids under electrostatic forces [65]. Between 1931 and 1944 Anton Formhals took out at least 22 patents on electrospinning [69]. In 1938, N.D. Rozenblum and I.V. Petryanov-Sokolov generated electrospun fibers, which they developed into filter materials [70]. Between 1964 and 1969 Sir Geoffrey Ingram Taylor produced the beginnings of a theoretical foundation of electrospinning by mathematically modeling the shape of the (Taylor) cone formed by the fluid droplet under the effect of an electric field [71, 72]. In the early 1990s several research groups (such as Reneker) demonstrated electrospun nano-fibers. Since 1995, the number of publications about electrospinning has been increasing exponentially every year [69].

1.3.1.5.2 ELECTROSPINNING PROCESS

Electrospinning process can be explained in five significant steps including [48, 73–75]:

a) **Charging of the polymer fluid**

The syringe is filled with a polymer solution, the polymer solution is charged with a very high potential around 10–30 kV. The nature of the fluid and polarity of the applied potential free electrons, ions or ion-pairs are generated as the charge carriers form an electrical double layer. This charging induction is suitable for conducting fluid, but for nonconducting fluid charge directly injected into the fluid by the application of electrostatic field.

b) **Formation of the cone jet (Taylor cone)**

The polarity of the fluid depends upon the voltage generator. The repulsion between the similar charges at the free electrical double layer works against the surface tension and fluid elasticity in the polymer solution to deform the droplet into a conical shaped structure i.e. known as a Taylor cone. Beyond a critical charge density Taylor-cone becomes unstable and a jet of fluid is ejected from the tip of the cone.

c) **Thinning of the jet in the presence of an electric field**

The jet travels a path to the ground; this fluid jet forms a slender continuous liquid filament. The charged fluid is accelerated in the presence of an electrical field. This region of fluid is generally linear and thin.

d) **Instability of the jet**

Fluid elements accelerated under electric field and thus stretched and succumbed to one or more fluid instabilities, which distort as they grow following many spiral and distort the path before collected on the collector electrode. This region of instability is also known as whipping region.

e) **Collection of the jet**

Charged electro spun fibers travel downfield until its impact with a lower potential collector plate. The orientation of the collector affects the alignment of the fibers. Different type of collector also affects the morphology and the properties of producing nanofiber. Different type of collectors are used: Rotating drum collector, moving belt collector, rotating wheel with beveled edge, multifilament thread, parallel bars, simple mesh collector, etc.

1.3.1.5.3 ELECTROSPINNING SET-UPS

Electrospinning is conducted at room temperature with atmospheric conditions. The typical set up of electrospinning apparatus is shown in Fig. 1.1. Basically, an electrospinning system consists of three major components: 1) a high voltage power supply, 2) a spinneret (such as a pipette tip) and 3) a grounded collecting plate (usually a metal screen, plate, or rotating mandrel); and uses a high voltage source to inject charge of a certain polarity into a polymer solution or melt, which is then accelerated towards a collector of opposite polarity [73, 76, 77]. Most of the polymers are dissolved in some solvents before electrospinning, and when it completely dissolves, forms polymer solution. The polymer fluid is then introduced into the capillary tube for electrospinning. However, some polymers may emit unpleasant or even harmful smells, so the processes should be conducted within chambers having a ventilation system. In the electrospinning process, a polymer solution held by its surface tension at the end of a capillary tube is subjected to an electric field and an electric charge is induced on the liquid surface due to this electric field. When the electric field applied reaches a critical value, the repulsive electrical forces overcome the surface tension forces. Eventually, a charged jet of the solution is ejected from the tip of the Taylor cone and an unstable and a rapid whipping of the jet occurs in the space between the capillary tip and collector which leads to evaporation of the solvent, leaving a polymer behind. The jet is only stable at the tip of the spinneret and after that instability starts. Thus, the electrospinning process offers a simplified technique for fiber formation [50, 73, 78, 79].

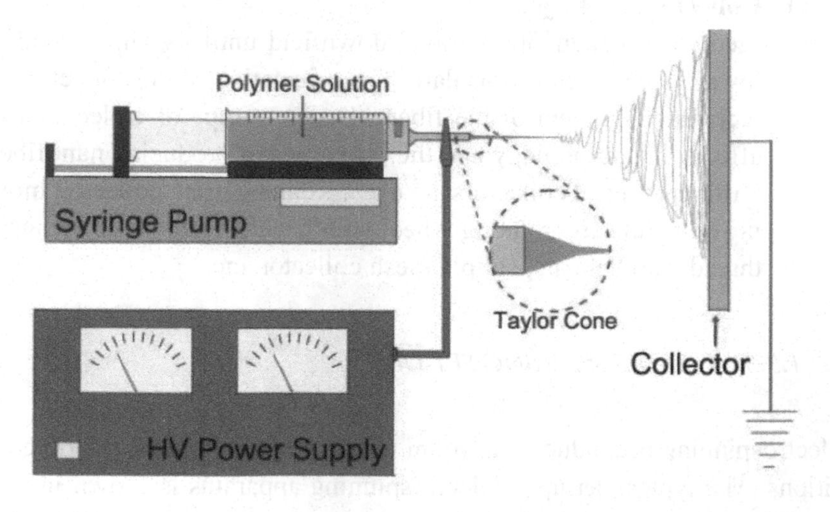

FIGURE 1.1 Scheme of a conventional electrospinning set-up.

1.3.1.5.4 THE EFFECTIVE PARAMETERS ON ELECTROSPINNING

The electrospinning process is generally governed by many parameters, which can be classified broadly into solution parameters, process parameters, and ambient parameters. Each of these parameters significantly affects the fiber morphology obtained as a result of electrospinning, and by proper manipulation of these parameters we can get nanofibers of desired morphology and diameters. These effective parameters are sorted as below [63, 67, 73, 76]:

a) **Polymer solution parameters,** which includes molecular weight and solution viscosity, surface tension, solution conductivity and dielectric effect of solvent.
b) **Processing parameters,** which include voltage, feed rate, temperature, effect of collector, the diameter of the orifice of the needle.

a) **Polymer solution parameters**

1) **Molecular weight and solution viscosity**

Higher the molecular weight of the polymer, increases molecular entanglement in the solution, hence there is an increase in viscosity. The electro spun jet eject with high viscosity during it is stretched to a collector electrode leading to formation of continuous fiber with higher diameter, but very high viscosity makes difficult to pump the solution and also lead to the drying of the solution at the needle tip. As a very low viscosity lead in bead formation in the resultant electro spun fiber, so the molecular weight and viscosity should be acceptable to form nanofiber [48, 80].

2) **Surface tension**

Lower viscosity leads to decrease in surface tension resulting bead formation along the fiber length because the surface area is decreased, but at the higher viscosity effect of surface tension is nullified because of the uniform distribution of the polymer solution over the entangled polymer molecules. So, lower surface tension is required to obtain smooth fiber and lower surface tension can be achieved by adding of surfactants in polymer solution [80, 81].

3) **Solution conductivity**

Higher conductivity of the solution followed a higher charge distribution on the electrospinning jet, which leads to increase in stretching of the solution during fiber formation. Increased conductivity of the polymer solution lowers the critical voltage for the electro spinning. Increased charge leads to the higher bending instability leading to the higher deposition area of the fiber being formed, as a result jet path is increased and finer fiber is formed. Solution conductivity can be increased by the addition of salt or polyelectrolyte or increased by the addition of drugs and proteins, which dissociate into ions when dissolved in the solvent formation of smaller diameter fiber [67, 80].

4) **Dielectric effect of solvent**

Higher the dielectric property of the solution lesser is the chance of bead formation and smaller is the diameter of electro-spun fiber. As the dielectric property is increased, there is increase in the bending instability of the jet and the deposition area of the fiber is increased. As jet path length is increased fine fiber deposit on the collector [67, 80].

b) **Processing condition parameters**

1) **Voltage**

Taylor cone stability depends on the applied voltage, at the higher voltage greater amount of charge causes the jet to accelerate faster leading to smaller and unstable Taylor cone. Higher voltage leads to greater stretching of the solution due to fiber with small diameter formed. At lower voltage the flight time of the fiber to a collector plate increases that led to the formation of fine fibers. There is greater tendency to bead formation at high voltage because of increased instability of the Taylor cone, and theses beads join to form thick diameter fibers. It is observed that the better crystallinity in the fiber obtained at higher voltage, because with very high voltage acceleration of fiber increased that reduced flight time and polymer molecules do not have much time to align them and fiber with less crystallinity formed. Instead of DC if AC voltage is provided for electro spinning it forms thicker fibers [48, 80].

2) **Feed rate**

As the feed rate is increased, there is an increase in the fiber diameter because greater volume of solution is drawn from the needle tip [80].

3) **Temperature**

At high temperature, the viscosity of the solution is decreased and there is increase in higher evaporation rate, which allows greater stretching of the solution and a uniform fiber is formed [82].

4) **Effect of collector**

In electro spinning, collector material should be conductive. The collector is grounded to create stable potential difference between needle and collector. A nonconducting material collector reduces the amount of fiber being deposited with lower packing density. But in case of conducting collector there is accumulation of closely packed fibers with higher packing density. Porous collector yields fibers with lower packing density as compared to nonporous collector plate. In porous collector plate the surface area is increased so residual solvent molecules gets evaporated fast as compared to nonporous. Rotating collector is useful in getting dry fibers as it provides more time to the solvents to evaporate. It also increases fiber morphology [83]. The specific hat target with proper parameters has a uniform surface electric field distribution, the target can collect the fiber mats of uniform thickness and thinner diameters with even size distribution [80].

5) **Diameter of pipette orifice**

Orifice with small diameter reduces the clogging effect due to less exposure of the solution to the atmosphere and leads to the formation of fibers with smaller diameter. However, very small orifice has the disadvantage that it creates problems in extruding droplets of solution from the tip of the orifice [80].

1.4 INTRODUCTION TO THEORETICAL STUDY OF ELECTROSPINNING PROCESS

Electrospinning is a procedure in which an electrical charge to draw very fine (typically on the micro or nano scale) fibers from polymer solution or molten. Electrospinning shares characteristics of both electrospraying and conventional solution dry spinning of fibers. The process does not require the use of coagulation chemistry or high temperatures to produce solid threads from solution. This makes the process more efficient to produce the fibers using large and complex molecules. Recently, various polymers have been successfully electrospun into ultrafine fibers mostly in solvent solution and some in melt form [79, 84]. Optimization of the alignment and morphology of the fibers is produced by fitting the composition of the solution and the configuration of the electrospinning apparatus such as voltage, flow rate, etc. As a result, the efficiency of this method can be improved [85]. Mathematical and theoretical modeling and simulating procedure will assist to offer an in-depth insight into the physical understanding of complex phenomena during electrospinningand might be very useful to manage contributing factors toward increasing production rate [75, 86].

Despite the simplicity of the electrospinning technology, industrial applications of it are still relatively rare, mainly due to the notable problems with very low fiber production rate and difficulties in controlling the process [67].

Modeling and simulation (M&S) give information about how something will act without actually testing it in real. The model is a representation of a real object or system of objects for purposes of visualizing its appearance or analyzing its behavior. Simulation is a transition from a mathematical or computational model for description of the system behavior based on sets of input parameters [87, 88]. Simulation is of ten the only means for accurately predicting the performance of the modeled system [89]. Using simulation is generally cheaper and safer than conducting experiments with a prototype of the final product. Also simulation can often be even more realistic than traditional experiments, as they allow the free configuration of environmental and operational parameters and can often be run faster than in real time. In a situation with different alternatives analysis, simulation can improve the efficiency, in particular when the necessary data to initialize can easily be obtained from operational data. Applying simulation adds decision support systems to the toolbox of traditional decision support systems [90].

Simulation permits set up a coherent synthetic environment that allows for integration of systems in the early analysis phase for a virtual test environment in the final system. If managed correctly, the environment can be migrated from the development and test domain to the training and education domain in real systems under realistic constraints [91].

A collection of experimental data and their confrontation with simple physical models appears as an effective approach towards the development of practical tools for controlling and optimizing the electrospinning process. On the other hand, it is necessary to develop theoretical and numerical models of electrospinning because of demanding a different optimization procedure for each material [92]. Utilizing a model to express the effect of electrospinning parameters will assist researchers to make an easy and systematic way of presenting the influence of variables and by means of that, the process can be controlled. Additionally, it causes to predict the results under a new combination of parameters. Therefore, without conducting any experiments, one can easily estimate features of the product under unknown conditions [93].

1.5 STUDY OF ELECTROSPINNING JET PATH

To yield individual fibers, most, if not all of the solvents must be evaporated by the time the electrospinning jet reaches the collection plate. As a result, volatile solvents are often used to dissolve the polymer. However, clogging of the polymer may occur when the solvent evaporates before the formation of the Taylor cone during the extrusion of the solution from several needles. In order to maintain a stable jet while still using a volatile solvent, an effective method is to use a gas jacket around the Taylor cone through two coaxial capillary tubes. The outer tube, which surrounds the inner tube will provide a controlled flow of inert gas, which is saturated with the solvent used to dissolve the polymer. The inner tube is then used to deliver the polymer solution. For 10-wt% poly (L-lactic acid) (PLLA) solution in dichloromethane, electrospinning was not possible due to clogging of the needle. However, when N2 gas was used to create a flowing gas jacket, a stable Taylor cone was formed and electrospinning was carried out smoothly.

1.5.1 THE THINNING JET (JET INSTABILITY)

The conical meniscus eventually gives rise to a slender jet that emerges from the apex of the meniscus and propagates downstream. Hohman et al.

[60] first reported this approach for the relatively simple case of Newtonian fluids. This suggests that the shape of the thinning jet depends significantly on the evolution of the surface charge density and the local electric field. As the jet thins down and the charges relax to the surface of the jet, the charge density and local field quickly pass through a maximum, and the current due to advection of surface charge begins to dominate over that due to bulk conduction.

The crossover occurs on the length scale given by [6]:

$$L_N = \left(K^4 Q^7 \rho^3 (\ln X)^2 / 8\pi^2 E_\infty I^5 \varepsilon^{-2}\right)^{1/5} \tag{1}$$

This length scale defines the "nozzle regime" over which the transition from the meniscus to the steady jet occurs. Sufficiently far from the nozzle regime, the jet thins less rapidly and finally enters the asymptotic regime, where all forces except inertial and electrostatic forces ceases to influence the jet. In this regime, the radius of the jet decreases as follows:

$$h = \left(\frac{Q^3 \rho}{2\pi^2 E_\infty I}\right)^{1/4} z^{-1/4} \tag{2}$$

Here z is the distance along the centerline of the jet. Between the "nozzle regime" and the "asymptotic regime," the evolution of the diameter of the thinning jet can be affected by the viscous response of the fluid. Indeed by balancing the viscous and the electrostatic terms in the force balance equation it can be shown that the diameter of the jet decreases as:

$$h = \left(\frac{6\mu Q^2}{\pi E_\infty I}\right)^{1/2} z^{-1} \tag{3}$$

In fact, the straight jet section has been studied extensively to understand the influence of viscoelastic behavior on the axisymmetric instabilities [94] and crystallization [60] and has even been used to extract extensional viscosity of polymeric fluids at very high strain rates.

For highly strain-hardening fluids, Yu et al. [95] demonstrated that the diameter of the jet decreased with a power-law exponent of −1/2, rather than −1/4 or −1, as discussed earlier for Newtonian fluids. This −1/2 power-law scaling for jet thinning in viscoelastic fluids has been explained in terms of a balance between electromechanical stresses acting on the surface of the jet and the viscoelastic stress associated with extensional strain hardening of the fluid. Additionally, theoretical studies of viscoelastic fluids predict a change in the shape of the jet due to non-Newtonian fluid behavior. Both Yu et al. [95] and Han et al. [96] have demonstrated that substantial elastic stresses can be accumulated in the fluid as a result of the high-strain rate in the transition from the meniscus into the jetting region. This elastic stress stabilizes the jet against external perturbations. Further downstream the rate of stretching slows down, and the longitudinal stresses relax through viscoelastic processes. The relaxation of stresses following an extensional deformation, such as those encountered in electrospinning, has been studied in isolation for viscoelastic fluids [97]. Interestingly, Yu et al. [95] also observed that, elastic behavior notwithstanding, the straight jet transitions into the whipping region when the jet diameter becomes of the order of 10 mm.

1.5.2 THE WHIPPING JET (JET INSTABILITY)

While it is in principle possible to draw out the fibers of small diameter by electrospinning in the cone-jet mode alone, the jet does not typically solidify enough en route to the collector and succumbs to the effect of force imbalances that lead to one or more types of instability. These instabilities distort the jet as they grow. A family of these instabilities exists, and can be analyzed in the context of various symmetries (axisymmetric or nonaxisymmetric) of the growing perturbation to the jet.

Some of the lower modes of this instability observed in electrospinning have been discussed in a separate review [81]. The "whipping instability" occurs when the jet becomes convectively unstable and its centerline bends. In this region, small perturbations grow exponentially, and the jet is stretched out laterally. Shin et al. [62] and Fridrikh et al. [63] have demonstrated how the whipping instability can be largely responsible for the formation of solid fiber in electrospinning. This is significant, since as recently as the late 1990s the bifurcation of the jet into two more or less equal parts (also called "splitting" or "splaying") were thought to be the mechanism through which the diameter of the jet is reduced, leading to the fine fibers formed in electrospinning. In contrast to "splitting" or "splaying," the appearances of secondary, smaller jets from the primary jet have been observed more frequently and in situ [64, 98]. These secondary jets occur when the conditions on the surface of the jet are such that perturbations in the local field, for example, due to the onset of the slight bending of the jet, is enough to overcome the surface tension forces and invoke a local jetting phenomenon.

The conditions necessary for the transition of the straight jet to the whipping jet has been discussed in the works of Ganan-Calvo [99], Yarin et al. [64], Reneker et al. [66] and Hohman et al. [60].

During this whipping instability, the surface charge repulsion, surface tension, and inertia were considered to have more influence on the jet path than Maxwell's stress, which arises due to the electric field and finite conductivity of the fluid. Using the equations reported by Hohman et al. [60, 61] and Fridrikh et al. [63] obtained an equation for the lateral growth of the jet excursions arising from the whipping instability far from the onset and deep into the nonlinear regime. These developments have been summarized in the review article of Rutledge and Fridrikh.

The whipping instability is postulated to impose the stretch necessary to draw out the jet into fine fibers. As discussed previously, the stretch imposed can make an elastic response in the fluid, especially if the fluid is polymeric in nature. An empirical rheological model was used to explore the consequences of nonlinear behavior of the fluid on the growth of the amplitude of the whipping instability in numerical calculations [63, 79]. There it was observed that the elasticity of the fluid significantly reduces the amplitude of oscillation of the whipping jet. The elastic response also stabilizes the jet against the effect of surface tension. In the absence of any elasticity, the jet eventually breaks up and forms an aerosol. However, the presence of a polymer in the fluid can stop this breakup if:

$$\tau / \left(\frac{\rho h^3}{\gamma} \right)^{1/2} \geq 1 \tag{4}$$

Where τ is the relaxation time of the polymer, ρ is the density of the fluid, h is a characteristic radius, and γ is the surface tension of the fluid.

1.6 ELECTROSPINNING DRAWBACKS

Electrospinning has attracted much attention both to academic research and industry application because electrospinning (1) can fabricate continuous fibers with diameters down to a few nanometers, (2) is applicable to a wide range of materials such as synthetic and natural polymers, metals as well as ceramics and composite systems, (3) can prepare nanofibers with low cost and high yielding [47].

Despite the simplicity of the electrospinning technology, industrial applications of it are still relatively rare, mainly due to the notable problems of very low fiber production rate and difficulties in controlling the process [50, 67]. The usual feed-rate for electrospinning is about 1.5 mL/hr. Given a solution concentration of 0.2 g/mL, the mass of nanofiber collected from a single needle after an hour is only 0.3 g. In order for electrospinning to be commercially viable, it is necessary to increase the production rate of the nanoflbers. To do so, multiple-spinning setup is necessary to increase the yield while at the same time maintaining the uniformity of the nanofiber mesh [48].

Optimization of the alignment and morphology of the fibers, which is produced by fitting the composition of the solution and the configuration of the electrospinning apparatus such as voltage, flow rate, etc., can be useful to improve the efficiency of this method [85]. Mathematical and theoretical modeling and simulating procedure will assist to offer an in-depth insight into the physical understanding of complex phenomena during electrospinningand might be very useful to managc contributing factors toward increasing production rate [75, 86].

Presently, nanofibers have attracted the attention of researchers due to their remarkable micro and nano structural characteristics, high surface area, small pore size, and the possibility of their producing three dimensional structure that enable the development of advanced materials with sophisticated applications [73].

Controlling the property, geometry, and mass production of the nanofibers, is essential to comprehend quantitatively how the electrospinning process transforms the fluid solution through a millimeter diameter capillary tube into solid fibers which are four to five orders smaller in diameter [74].

As mentioned above, the electrospinning gives us the impression of being a very simple and easily controlled technique for the production of nanofibers. But, actually the process is very intricate. Thus, electrospinning is usually described as the interaction of several physical instability processes. The bending and stretching of the jet are mainly caused by the rapidly whipping which is an essential element of the process induced by these instabilities. Until now, little is known about the detailed mechanisms of the instabilities and the splaying process of the primary jet into multiple filaments. It is thought to be responsible that the electrostatic forces overcome surface tensions of the droplet during undergoing the electrostatic field and the deposition of jets formed nanofibers [47].

Though electrospinning has been become an indispensable technique for generating functional nanostructures, many technical issues still need to be resolved. For example, it is not easy to prepare nanofibers with a same scale in diameters by electrospinning; it is still necessary to investigate systematically the correlation between the secondary structure of nanofiber and the processing parameters; the mechanical properties, photoelectric properties and other property of single fiber should be systematically studied and optimized; the production of nanofiber is still in laboratory level, and it is extremely important to make efforts for scaled-up commercialization; nanofiber from electrospinning has a the low production rate and low mechanical strength which hindered it's commercialization; in addition, another more important issue should be resolved is how to treat the solvents volatilized in the process.

Until now, lots of efforts are putted on the improvement of electrospinning installation, such as the shape of collectors, modified spinnerets and so on. The application of multi-jets electrospinning provides a possibility to produce nanofibers in industrial scale. The development of equipments, which can collect the poisonous solvents and the application of melt electrospinning, which would largely reduce the environment problem, create a possibility of the industrialization of electrospinning. The application of water as the solvent for elelctrospinning provides another approach to reduce environmental pollution, which is the main fact hindered the commercialization of electrospinning. In summary, electrospinning is an attractive and promising approach for the preparation of functional nanofibers due to its wide applicability to materials, low cost and high production rate [47].

1.7 MODELLING OF THE ELECTROSPINNING PROCESS

The electrospinning process is a fluid dynamics related problem. Controlling the property, geometry, and mass production of the nanofibers, is essential to comprehend quantitatively how the electrospinning process transforms the fluid solution through a millimeter diameter capillary tube into solid fibers which are four to five orders smaller in diameter [74]. Although information on the effect of various processing parameters and constituent material properties can be obtained experimentally, theoretical models offer in-depth scientific understanding which can be useful to clarify the affecting factors that cannot be exactly measured experimentally. Results from modeling also explained how processing parameters and fluid behavior lead to the nanofiber of appropriate properties. The term "properties" refers to basic properties (such as, fiber diameter, surface roughness, fiber connectivity, etc.), physical properties (such as, stiffness, toughness, thermal conductivity, electrical resistivity, thermal expansion coefficient, density, etc.) and specialized properties (such as, biocompatibility, degradation curve, etc. for biomedical applications) [48, 73].

For example, the developed models can be used for the analysis of mechanisms of jet deposition and alignment on various collecting devices in arbitrary electric fields [100].

The various method formulated by researchers are prompted by several applications of nanofibers. It would be sufficient to briefly describe some of these methods to observed similarities and disadvantages of these approaches. An abbreviated literature review of these models will be discussed in the following sections.

1.7.1 ASSUMPTIONS

Just as in any other process modeling, a set of assumptions are required for the following reasons:

a) to furnish industry-based applications whereby speed of calculation, but not accuracy, is critical;
b) to simplify and therefore, enabling checkpoints to be made before more detailed models can proceed; and

c) for enabling the formulations to be practically traceable.

The first assumption to be considered as far as electrospinning is concerned is conceptualizing the jet itself. Even though the most appropriate view of a jet flow is that of a liquid continuum, the use of nodes connected in series by certain elements that constitute rheological properties has proven successful [64, 66]. The second assumption is the fluid constitutive properties. In the discrete node model [66], the nodes are connected in series by a Maxwell unit, that is, a spring and dashpot in series, for quantifying the viscoelastic properties.

In analyzing viscoelastic models, we apply two types of elements: the dashpot element, which describes the force as being in proportion to the velocity (recall friction), and the spring element, which describes the force as being in proportion to elongation. One can then develop viscoelastic models using combinations of these elements. Among all possible viscoelastic models, the Maxwell model was selected by [66] due to its suitability for liquid jet as well as its simplicity. Other models are either unsuitable for liquid jet or too detailed.

In the continuum models a power law can be used for describing the liquid behavior under shear flow for describing the jet flow [101]. At this juncture, we note that the power law is characterized from a shear flow, while the jet flow in electrospinning undergoes elongational flow. This assumption will be discussed in detail in following sections.

The other assumption that should be applied in electrospinning modeling is about the coordinate system. The method for coordinate system selection in electrospinning process is similar to other process modeling, the system that best bring out the results by (i) allowing the computation to be performed in the most convenient manner and, more importantly, (ii) enabling the computation results to be accurate. In view of the linear jet portion during the initial first stage of the jet, the spherical coordinate system is eliminated. Assuming the second stage of the jet to be perfectly spiraling, due to bending of the jet, the cylindrical coordinate system would be ideal. However, experimental results have shown that the bending instability portion of the jet is not perfectly expanding spiral. Hence the Cartesian coordinate system, which is the most common of all coordinate system, is adopted.

Depending on the processing parameters (such as applied voltage, volume flow rate, etc.) and the fluid properties (such as surface tension, viscosity, etc.) as many as 10 modes of electrohydrodynamically driven liquid jet have been identified [102]. The scope of jet modes is highly abbreviated in this chapter because most electrospinning processes that lead to nanofibers consist of only two modes, the straight jet portion and the spiraling (or whipping) jet portion. Insofar as electrospinning process modeling is involved, the following classification indicates the considered modes or portion of the electrospinning jet.

1. Modeling the criteria for jet initiation from the droplet [64, 103];
2. Modeling the straight jet portion [104, 105] Spivak et al. [101, 106];
3. Modeling the entire jet [60, 61, 66, 107].

A schematic of the jet flow variety, which occurs in electrospinning process presented in Fig. 1.2.

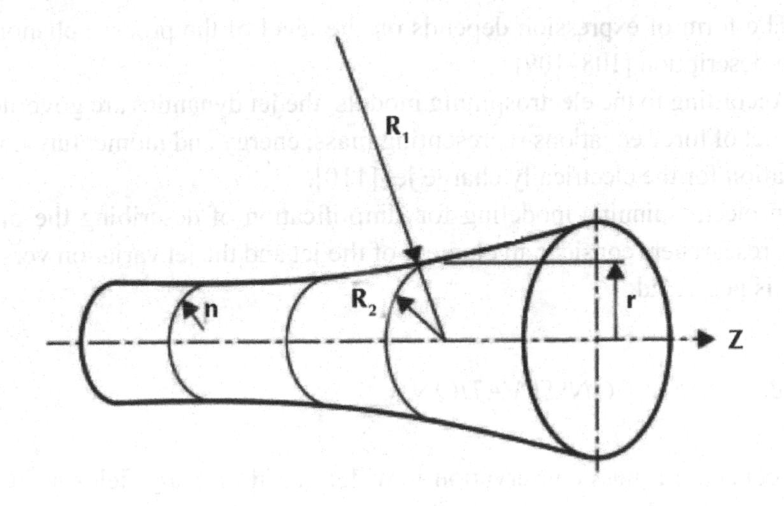

FIGURE 1.2 Geometry of the jet flow.

1.7.2 CONSERVATION RELATIONS

Balance of the producing accumulation is, particularly, a basic source of quantitative models of phenomena or processes. Differential balance equations are formulated for momentum, mass and energy through the contribution of local rates of transport expressed by the principle of Newton's, Fick's and Fourier laws. For a description of more complex systems like electrospinning that involved strong turbulence of the fluid flow, characterization of the product property is necessary and various balances are required [108].

The basic principle used in modeling of chemical engineering process is a concept of balance of momentum, mass and energy, which can be expressed in a general form as:

$$A = I + G - O - C \tag{5}$$

where, A is the accumulation built up within the system; I is the input entering through the system surface; G is the generation produced in system volume; O is the output leaving through system boundary; and C is the consumption used in system volume.

The form of expression depends on the level of the process phenomenon description [108–109]

According to the electrospinnig models, the jet dynamics are governed by a set of three equations representing mass, energy and momentum conservation for the electrically charge jet [110].

In electrospinning modeling for simplification of describing the process, researchers consider an element of the jet and the jet variation versus time is neglected.

1.7.2.1 MASS CONSERVATION

The concept of mass conservation is widely used in many fields such as chemistry, mechanics, and fluid dynamics. Historically, mass conservation was discovered in chemical reactions by Antoine Lavoisier in the late eighteenth century, and was of decisive importance in the progress from alchemy to the modern natural science of chemistry. The concept of matter conservation is useful and sufficiently accurate for most chemical calculations, even in modern practice [111].

The equations for the jet follow from Newton's Law and the conservation laws obey, namely, conservation of mass and conservation of charge [60].

According to the conservation of mass equation,

$$\pi R^2 \upsilon = Q \tag{6}$$

$$\frac{\partial}{\partial t}\left(\pi R^2\right) + \frac{\partial}{\partial z}\left(\pi R^2 \upsilon\right) = 0 \tag{7}$$

For incompressible jets, by increasing the velocity the radius of the jet decreases. At the maximum level of the velocity, the radius of the jet reduces. The macromolecules of the polymers are compacted together closer while the jet becomes thinner as it shown in Fig. 1.3. When the radius of the jet reaches the minimum value and its speed becomes maximum to keep the conservation of mass equation, the jet dilates by decreasing its density, which called electrospinning dilation [112–113].

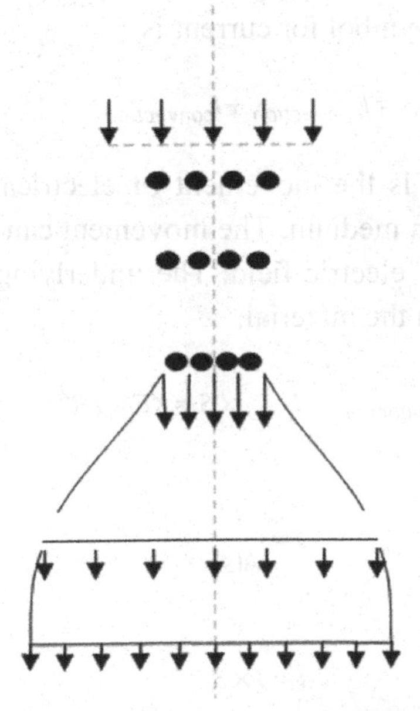

FIGURE 1.3 Macromolecular chains are compacted during the electrospinning.

1.7.2.2 ELECTRIC CHARGE CONSERVATION

An electric current is a flow of electric charge. Electric charge flows when there is voltage present across a conductor. In physics, charge conservation is the principle that electric charge can neither be created nor destroyed. The net quantity of electric charge, the amount of positive charge minus the amount of negative charge in the universe, is always conserved. The first written statement of the principle was by American scientist and statesman Benjamin Franklin in 1747 [114]. Charge conservation is a physical law, which states that the change in the amount of electric charge in any volume of space is exactly equal to the amount of charge in a region and the flow of charge into and out of that region [115].

During the electrospinning process, the electrostatic repulsion between excess charges in the solution stretches the jet. This stretching also decreases the jet diameter that this leads to the law of charge conservation as the second governing equation [116].

In electrospinning process, the electric current, which induced by electric field, is included into two parts, conduction and convection.

The conventional symbol for current is I:

$$I = I_{conduction} + I_{convection} \tag{8}$$

Electrical conduction is the movement of electrically charged particles through a transmission medium. The movement can form an electric current in response to an electric field. The underlying mechanism for this movement depends on the material.

$$I_{conduction} = J_{cond} \times S = KE \times \pi R^2 \tag{9}$$

$$J = \frac{I}{A(s)} \tag{10}$$

$$I = J \times S \tag{11}$$

Convection current is the flow of current with the absence of an electric field.

$$I_{convection} = J_{conv} \times S = 2\pi R(L) \times \sigma v \quad (12)$$

$$J_{conv} = \sigma v \quad (13)$$

So, the total current can be calculated as:

$$\pi R^2 KE + 2\pi R v \sigma = I \quad (14)$$

$$\frac{\partial}{\partial t}(2\pi R\sigma) + \frac{\partial}{\partial z}\left(\pi R^2 KE + 2\pi R v \sigma\right) = 0 \quad (15)$$

1.7.2.3 MOMENTUM BALANCE

In classical mechanics, linear momentum or translational momentum is the product of the mass and velocity of an object. Like velocity, linear momentum is a vector quantity, possessing a direction as well as a magnitude:

$$P = m\upsilon \quad (16)$$

Linear momentum is also a conserved quantity, meaning that if a closed system (one that does not exchange any matter with the outside and is not acted on by outside forces) is not affected by external forces, its total linear momentum cannot change. In classical mechanics, conservation of linear momentum is implied by Newton's laws of motion; but it also holds in special relativity (with a modified formula) and, with appropriate definitions, a (generalized) linear momentum conservation law holds in electrodynamics, quantum mechanics, quantum field theory, and general relativity [114]. For example, according to the third law, the forces between two particles are equal and opposite. If the particles are numbered 1 and 2, the second law states:

$$F_1 = \frac{dP_1}{dt} \quad (17)$$

$$F_2 = \frac{dP_2}{dt} \quad (18)$$

Therefore:

$$\frac{dP_1}{dt} = -\frac{dP_2}{dt} \quad (19)$$

$$\frac{d}{dt}(P_1 + P_2) = 0 \quad (20)$$

If the velocities of the particles are υ_{11} and υ_{12} before the interaction, and afterwards they are υ_{21} and υ_{22}, then,

$$m_1\upsilon_{11} + m_2\upsilon_{12} = m_1\upsilon_{21} + m_2\upsilon_{22} \quad (21)$$

This law holds no matter how complicated the force is between the particles. Similarly, if there are several particles, the momentum exchanged between each pair of particles adds up to zero, so the total change in momentum is zero. This conservation law applies to all interactions, including collisions and separations caused by explosive forces. It can also be generalized to situations where Newton's laws do not hold, for example in the theory of relativity and in electrodynamics [104, 117]. The momentum equation for the fluid can be derived as follow:

$$\rho(\frac{d\upsilon}{dt} + \upsilon\frac{d\upsilon}{dz}) = \rho g + \frac{d}{dz}[\tau_{zz} - \tau_{rr}] + \frac{\gamma}{R^2}.\frac{dr}{dz} + \frac{\sigma}{\varepsilon_0}\frac{d\sigma}{dz} + (\varepsilon - \varepsilon_0)(E\frac{dE}{dz}) + \frac{2\sigma E}{r} \quad (22)$$

But commonly the momentum equation for electrospinning modeling is formulated by considering the forces on a short segment of the jet [104, 117].

$$\frac{d}{dz}(\pi R^2 \rho \upsilon^2) = \pi R^2 \rho g + \frac{d}{dz}\left[\pi R^2(-p + \tau_{zz})\right] + \frac{\gamma}{R}.2\pi RR' + 2\pi R(t_t^e - t_n^e R') \quad (23)$$

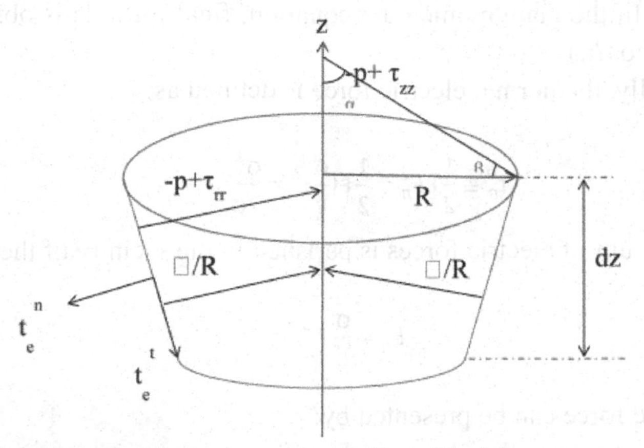

FIGURE 1.4 Momentum balance on a short section of the jet.

As it is shown in the Fig. 1.4. the element's angels could be defined as α and β. According to the mathematical relationships, it is obvious that:

$$\alpha+\beta=\frac{\pi}{2} \tag{24}$$

$$\begin{aligned} \sin\alpha &= \tan\alpha \\ \cos\alpha &= 1 \end{aligned} \tag{25}$$

Due to the Fig. 1.4, relationships between these electrical forces are as below:

$$t_n^e \sin\alpha \cong t_n^e \tan\alpha \cong -t_n^e \tan\beta \cong -\frac{dR}{dz}t_n^e = -R' t_n^e \tag{26}$$

$$t_t^e \cos\alpha \cong t_t^e \tag{27}$$

So the effect of the electric forces in the momentum balance equation can be presented as:

$$2\pi RL(t_t^e - R' t_n^e)dz \tag{28}$$

(Notation: In the main momentum equation, final formula is obtained by dividing into *dz*.)

Generally, the normal electric force is defined as:

$$t_n^e \cong \frac{1}{2}\bar{\varepsilon}E_n^2 = \frac{1}{2}\bar{\varepsilon}(\frac{\sigma}{\bar{\varepsilon}})^2 = \frac{\sigma^2}{2\bar{\varepsilon}} \tag{29}$$

A little amount of electric forces is perished in the vicinity of the air.

$$E_n = \frac{\sigma}{\bar{\varepsilon}} \tag{30}$$

The electric force can be presented by:

$$F = \frac{\Delta We}{\Delta l} = \frac{1}{2}(\varepsilon - \bar{\varepsilon})E^2 \times \Delta S \tag{31}$$

The force per surface unit is:

$$\frac{F}{\Delta S} = \frac{1}{2}(\varepsilon - \bar{\varepsilon})E^2 \tag{32}$$

Generally, the electric potential energy is obtained by:

$$Ue = -We = -\int F.ds \tag{33}$$

$$\Delta We = \frac{1}{2}(\varepsilon - \bar{\varepsilon})E^2 \times \Delta V = \frac{1}{2}(\varepsilon - \bar{\varepsilon})E^2 \times \Delta S.\Delta l \tag{34}$$

So, finally it could be resulted:

$$t_n^e = \frac{\sigma^2}{2\bar{\varepsilon}} - \frac{1}{2}(\varepsilon - \bar{\varepsilon})E^2 \tag{35}$$

$$t_t^e = \sigma E \tag{36}$$

1.7.2.4 COULOMB'S LAW

Coulomb's law is a mathematical description of the electric force between charged objects, which is formulated by the 18th-century French physicist Charles-Augustin de Coulomb. It is analogous to Isaac Newton's law of gravity. Both gravitational and electric forces decrease with the square of the distance between the objects, and both forces act along a line between them [118]. In Coulomb's law, the magnitude and sign of the electric force are determined by the electric charge, more than the mass of an object. Thus, a charge, which is a basic property matter, determines how electromagnetism affects the motion of charged targets [114].

Coulomb force is thought to be the main cause for the instability of the jet in the electrospinning process [119]. This statement is based on the Earnshaw's theorem, named after Samuel Earnshaw [120], which claims that "A charged body placed in an electric field of force cannot rest in stable equilibrium under the influence of the electric forces alone." This theorem can be notably adapted to the electrospinning process [119]. The instability of charged jet influences on jet deposition and as a consequence on nanofiber formation. Therefore, some researchers applied developed models to the analysis of mechanisms of jet deposition and alignment on various collecting devices in arbitrary electric fields [66].

The equation for the potential along the centerline of the jet can be derived from Coulomb's law. Polarized charge density is obtained:

$$\rho_{p'} = -\vec{\nabla}.\vec{P}' \tag{37}$$

Where P' is polarization:

$$\vec{P}' = (\varepsilon - \overline{\varepsilon})\vec{E} \tag{38}$$

By substituting P' in Eq. (38):

$$\rho_{P'} = -(\overline{\varepsilon} - \varepsilon)\frac{dE}{dz'} \tag{39}$$

Beneficial charge per surface unit can be calculated as below:

$$\rho_{P'} = \frac{Q_b}{\pi R^2} \tag{40}$$

$$Q_b = \rho_b . \pi R^2 = -(\overline{\varepsilon} - \varepsilon)\pi R^2 \frac{dE}{dz'} \tag{41}$$

$$Q_b = -(\overline{\varepsilon} - \varepsilon)\pi \frac{d(ER^2)}{dz'} \tag{42}$$

$$\rho_{sb} = Q_b .dz' = -(\overline{\varepsilon} - \varepsilon)\pi \frac{d}{dz'}(ER^2)dz' \tag{43}$$

The main equation of Coulomb's law:

$$F = \frac{1}{4\pi\varepsilon_0} \frac{qq_0}{r^2} \tag{44}$$

The electric field is:

$$E = \frac{1}{4\pi\varepsilon_0} \frac{q}{r^2} \tag{45}$$

The electric potential can be measured:

$$\Delta V = -\int E.dL \tag{46}$$

$$V = \frac{1}{4\pi\varepsilon_0} \frac{Q_b}{r} \tag{47}$$

According to the beneficial charge equation, the electric potential could be rewritten as:

$$\Delta V = Q(z) - Q_\infty(z) = \frac{1}{4\pi\overline{\varepsilon}} \int \frac{(q - Q_b)}{r} dz' \tag{48}$$

$$Q(z) = Q_\infty(z) + \frac{1}{4\pi\overline{\varepsilon}} \int \frac{q}{r} dz' - \frac{1}{4\pi\overline{\varepsilon}} \int \frac{Q_b}{r} dz' \tag{49}$$

$$Q_b = -(\overline{\varepsilon} - \varepsilon)\pi \frac{d(ER^2)}{dz'} \tag{50}$$

The surface charge density's equation is:

$$q = \sigma.2\pi RL \tag{51}$$

$$r^2 = R^2 + (z - z')^2 \tag{52}$$

$$r = \sqrt{R^2 + (z - z')^2} \tag{53}$$

The final equation, which obtained by substituting the mentioned equations is:

$$Q(z) = Q_\infty(z) + \frac{1}{4\pi\overline{\varepsilon}} \int \frac{\sigma.2\pi R}{\sqrt{(z - z')^2 + R^2}} dz' - \frac{1}{4\pi\overline{\varepsilon}} \int \frac{(\overline{\varepsilon} - \varepsilon)\pi}{\sqrt{(z - z')^2 + R^2}} \frac{d(ER^2)}{dz'} \tag{54}$$

It is assumed that β is defined:

$$\beta = \frac{\varepsilon}{\overline{\varepsilon}} - 1 = -\frac{(\overline{\varepsilon} - \varepsilon)}{\overline{\varepsilon}} \tag{55}$$

So, the potential equation becomes:

$$Q(z) = Q_\infty(z) + \frac{1}{2\overline{\varepsilon}} \int \frac{\sigma.R}{\sqrt{(z - z')^2 + R^2}} dz' - \frac{\beta}{4} \int \frac{1}{\sqrt{(z - z')^2 + R^2}} \frac{d(ER^2)}{dz'} \tag{56}$$

The asymptotic approximation of χ is used to evaluate the integrals mentioned above:

$$\chi = \left(-z + \xi + \sqrt{z^2 - 2z\xi + \xi^2 + R^2}\right) \tag{57}$$

where c is "aspect ratio" of the jet (L = length, R0 = Initial radius).This leads to the final relation to the axial electric field:

$$E(z) = E_{\infty}(z) - \ln\chi\left(\frac{1}{\varepsilon}\frac{d(\sigma R)}{dz} - \frac{\beta}{2}\frac{d^2(ER^2)}{dz^2}\right) \tag{58}$$

1.7.2.5 FORCES CONSERVATION

There exists a force, as a result of charge build-up, acting upon the droplet coming out of the syringe needle pointing toward the collecting plate, which can be either grounded or oppositely charged. Furthermore, similar charges within the droplet promote jet initiation due to their repulsive forces. Nevertheless, surface tension and other hydrostatic forces inhibit the jet initiation because the total energy of a droplet is lower than that of a thin jet of equal volume upon consideration of surface energy. When the forces that aid jet initiation (such as electric field and Coulombic) overcome the opposing forces (such as surface tension and gravitational), the droplet accelerates toward the collecting plate. This forms a jet of very small diameter. Other than initiating jet flow, the electric field and Coulombic forces tend to stretch the jet, thereby contributing towards the thinning effect of the resulting nanofibers.

In the flow path modeling, we recall the Newton's Second Law of motion,

$$m\frac{d^2P}{dt^2} = \Sigma f \tag{59}$$

where, m (equivalent mass) and the various forces are summed as,

$$\Sigma f = f_C + f_E + f_V + f_S + f_A + f_G + \ldots \tag{60}$$

where, subscripts C, E, V, S, A and G correspond to the Coulombic, electric field, viscoelastic, surface tension, air drag and gravitational forces respectively. A description of each of these forces based on the literature [66] is summarized in Table 1.1. Here, V_0 = applied voltage; h = distance from pendent drop to ground collector; σ_V = viscoelastic stress; and ν = kinematic viscosity.

TABLE 1.1 Description of itemized forces or terms related to them.

Forces	Equations
Coulombic	$f_C = \frac{q^2}{l^2}$
Electric field	$f_E = -\frac{qV_0}{h}$
Viscoelastic	$f_V = \frac{d\sigma_V}{dt} = \frac{G}{l}\frac{dl}{dt} - \frac{G}{\eta}\sigma_V$
Surface tension	$f_S = \frac{\alpha\pi R^2 k}{\sqrt{x_i^2 + y_i^2}}\left[i\lvert x\rvert Sin(x) + i\lvert y\rvert Sin(y)\right]$
Air drag	$f_A = 0.65\pi R\rho_{air} v^2 \left(\frac{2vR}{v_{air}}\right)^{-0.81}$
Gravitational	$f_G = \rho g \pi R^2$

1.7.3 CONSTITUTIVE EQUATIONS

In modern condensed matter physics, the constitutive equation plays a major role. In physics and engineering, a constitutive equation or relation is a relation between two physical quantities that is specific to a material or substance, and approximates the response of that material to external stimulus, usually as applied fields or forces [121]. There are a sort of mechanical equation of state, and describe how the material is constituted mechanically. With these constitutive relations, the vital role of the material is reasserted [122]. There are two groups of constitutive equations: Linear and nonlinear constitutive equations [123]. These equations are combined with other governing physical laws to solve problems; for example in fluid mechanics the flow of a fluid in a pipe, in solid state physics the response of a crystal to an electric field, or in structural analysis, the connection between applied stresses or forces to strains or deformations [121].

The first constitutive equation (constitutive law) was developed by Robert Hooke and is known as Hooke's law. It deals with the case of linear elastic materials. Following this discovery, this type of equation, often called a "stress-strain relation" in this example, but also called a "constitutive assumption" or an "equation of state" was commonly used [124]. Walter Noll advanced the use of constitutive equations, clarifying their classification and the role of invariance requirements, constraints, and definitions of terms like "material," "isotropic," "aeolotropic," etc. The class of "constitutive relations" of the form stress rate = f (velocity gradient, stress, density) was the subject of Walter Noll's dissertation in 1954 under Clifford Truesdell [121]. There are several kinds of constitutive equations, which are applied commonly in electrospinning. Some of these applicable equations are discussed as following:

1.7.3.1 OSTWALD–DE WAELE POWER LAW

Rheological behavior of many polymer fluids can be described by power law constitutive equations [123]. The equations that describe the dynamics in electrospinning constitute, at a minimum, those describing the conservation of mass, momentum and charge, and the electric field equation. Additionally, a constitutive equation for the fluid behavior is also required [76]. A Power-law fluid, or the Ostwald–de Waele relationship, is a type of generalized Newtonian fluid for which the shear stress, τ, is given by:

$$\tau = K'\left(\frac{\partial \upsilon}{\partial y}\right)^m \tag{61}$$

which $\partial v/\partial y$ is the shear rate or the velocity gradient perpendicular to the plane of shear. The power law is only a good description of fluid behavior across the range of shear rates to which the coefficients are fitted. There are a number of other models that better describe the entire flow behavior of shear-dependent fluids, but they do so at the expense of simplicity, so the power law is still used to describe fluid behavior, permit mathematical predictions, and correlate experimental data [117, 125].

Nonlinear rheological constitutive equations applicable for polymer fluids (Ostwald–de Waele power law) were applied to the electrospinning process by Spivak and Dzenis [77, 101, 126].

$$\hat{\tau}^c = \mu\left[tr\left(\dot{\hat{\gamma}}^2\right)\right]^{(m-1)/2} \dot{\hat{\gamma}} \tag{62}$$

$$\mu = K\left(\frac{\partial \upsilon}{\partial y}\right)^{m-1} \tag{63}$$

Viscous Newtonian fluids are described by a special case of equation above with the flow index $m = 1$. Pseudoplastic (shear thinning) fluids are described by flow indices $0 \leq m \leq 1$. Dilatant (shear thickening) fluids are described by the flow indices $m > 1$ [101].

1.7.3.2 GIESEKUS EQUATION

In 1966, Giesekus established the concept of anisotropic forces and motions into polymer kinetic theory. With particular choices for the tensors describing the anisotropy, one can obtained Giesekus constitutive equation from elastic dumbbell kinetic theory [127, 128]. The Giesekus equation is known to predict, both qualitatively and quantitavely, material functions for steady and nonsteady shear and elongational flows. However, the equation sustains two drawbacks: it predicts that the viscosity is inversely proportional to the shear rate in the limit of infinite shear rate and it is unable to predict any decrease in the elongational viscosity with increasing elongation rates in uniaxial elongational flow. The first one is not serious because of the retardation time, which is included in the constitutive equation but the second one is more critical because the elongational viscosity of some polymers decreases with increasing of elongation rate [129, 130].

In the main Giesekus equation, the tensor of excess stresses depending on the motion of polymer units relative to their surroundings was connected to a sequence of tensors characterizing the configurational state of the different kinds of network structures present in the concentrated solution or melt. The respective set of constitutive equations indicates [131, 132]:

$$S_k + \eta \frac{\partial C_k}{\partial t} = 0 \quad (64)$$

The Eq. (65) indicates the upper convected time derivative (Oldroyd derivative):

$$\frac{\partial C_k}{\partial t} = \frac{DC_k}{Dt} - \left[C_k \nabla \upsilon + (\nabla \upsilon)^T C_k \right] \quad (65)$$

(Note: The upper convective derivative is the rate of change of any tensor property of a small parcel of fluid that is written in the coordinate system rotating and stretching with the fluid.)

C_k also can be measured as follows:

$$C_k = 1 + 2E_k \quad (66)$$

According to the concept of "recoverable strain" S_k may be understood as a function of E_k and vice versa. If linear relations corresponding to Hooke's law are adopted.

$$S_k = 2\mu_k E_k \quad (67)$$

So,

$$S_k = \mu_k (C_k - 1) \quad (68)$$

The Eq. (55) becomes:

$$S_k + \lambda_k \frac{\partial S_k}{\partial t} = 2\eta D \quad (69)$$

$$\lambda_k = \frac{\eta}{\mu_k} \quad (70)$$

As a second step in order to rid the model of the shortcomings is the scalar mobility constants B_k, which are contained in the constants η. This mobility constant can be represented as:

$$\tfrac{1}{2}(\beta_k S_k + S_k \beta_k) + \breve{\eta}\frac{\partial C_k}{\partial t} = 0 \tag{71}$$

The two parts of Eq. (62) reduces to the single constitutive equation:

$$\beta_k + \breve{\eta}\frac{\partial C_k}{\partial t} = 0 \tag{72}$$

The excess tension tensor in the deformed network structure where the well-known constitutive equation of a so-called Neo-Hookean material is proposed [131, 133]:

Neo-Hookean equation:

$$S_k = 2\mu_k E_k = \mu_k (C_k - 1) \tag{73}$$

$$\mu_k = NKT$$

$$\beta_k = 1 + \alpha(C_k - 1) = (1 - \alpha) + \alpha C_k \tag{74}$$

where K is Boltzmann's constant.

By substitution Eqs. (64) and (65) in the Eq. (61), it can obtained where the condition $0 \leq \alpha \leq 1$ must be fulfilled, the limiting case α=0 corresponds to an isotropic mobility [134].

$$0 \leq \alpha \leq 1 \quad [1 + \alpha(C_k - 1)](C_k - 1) + \lambda_k \frac{\partial C_k}{\partial t} = 0 \tag{75}$$

$$\alpha = 1 \quad C_k(C_k - 1) + \lambda_k \frac{\partial C_k}{\partial t} = 0 \tag{76}$$

$$0 \leq \alpha \leq 1 \quad C_k = \frac{S_k}{\mu_k} + 1 \tag{77}$$

By substituting equations above in Eq. (61), it becomes:

$$\left[1+\frac{\alpha S_k}{\mu_k}\right]\frac{S_k}{\mu_k}+\lambda_k\frac{\partial C_k}{\partial t}=0 \tag{78}$$

$$\frac{S_k}{\mu_k}+\frac{\alpha S_k^2}{\mu_k^2}+\lambda_k\frac{\partial\left(S_k/\mu_k+1\right)}{\partial t}=0 \tag{79}$$

$$\frac{S_k}{\mu_k}+\frac{\alpha S_k^2}{\mu_k^2}+\frac{\lambda_k}{\mu_k}\frac{\partial S_k}{\partial t}=0 \tag{80}$$

$$S_k+\frac{\alpha S_k^2}{\mu_k}+\lambda_k\frac{\partial S_k}{\partial t}=0 \tag{81}$$

D means the rate of strain tensor of the material continuum [131].

$$D=\frac{1}{2}\left[\nabla\upsilon+(\nabla\upsilon)^T\right] \tag{82}$$

The equation of the upper convected time derivative for all fluid properties can be calculated as:

$$\frac{\partial\otimes}{\partial t}=\frac{D\otimes}{Dt}-\left[\otimes.\nabla\upsilon+(\nabla\upsilon)^T.\otimes\right] \tag{83}$$

$$\frac{D\otimes}{Dt}=\frac{\partial\otimes}{\partial t}+\left[(\upsilon.\nabla).\otimes\right] \tag{84}$$

By replacing S_k instead of the symbol:

$$\lambda_k\frac{\partial S_k}{\partial t}=\lambda_k\frac{DS_k}{Dt}-\lambda_k\left[S_k\nabla\upsilon+(\nabla\upsilon)^T S_k\right]=\lambda_k\frac{DS_k}{Dt}-\lambda_k(\upsilon.\nabla)S_k \tag{85}$$

By simplification the equation above:

$$S_k+\frac{\alpha S_k^2}{\mu_k}+\lambda_k\frac{DS_k}{Dt}=\lambda_k(\upsilon.\nabla)S_k \tag{86}$$

$$S_k = 2\mu_k E_k \tag{87}$$

The assumption of $E_k = 1$ would lead to Eq. (88):

$$S_k + \frac{\alpha\lambda_k S_k^2}{\eta} + \lambda_k \frac{DS_k}{Dt} = \frac{\eta}{\mu_k}(2\mu_k)D = 2\eta D = \eta\left[\nabla\upsilon + (\nabla\upsilon)^T\right] \tag{88}$$

In electrospinning modeling articles τ is used commonly instead of S_k [105, 110, 112].

$$S_k \leftrightarrow \tau$$

$$\tau + \frac{\alpha\lambda_k \tau^2}{\eta} + \lambda_k \tau_{(1)} = \eta\left[\nabla\upsilon + (\nabla\upsilon)^T\right] \tag{89}$$

1.7.3.3 MAXWELL EQUATION

Maxwell's equations are a set of partial differential equations that, together with the Lorentz force law, form the foundation of classical electrodynamics, classical optics, and electric circuits. These fields are the bases of modern electrical and communications technologies. Maxwell's equations describe how electric and magnetic fields are generated and altered by each other and by charges and currents. They are named after the Scottish physicist and mathematician James Clerk Maxwell who published an early form of those equations between 1861 and 1862 [135, 136]. It will be discussed in the next section in detail.

1.7.4 MICROSCOPIC MODELS

One of the aims of computer simulation is to reproduce experiment to elucidate the invisible microscopic details and further explain the experiments. Physical phenomena occurring in complex materials cannot be encapsulated within a single numerical paradigm. In fact, they should be described within hierarchical, multilevel numerical models in which each submodel is responsible for different spatial-temporal behavior and passes out the averaged parameters of the model, which is next in the hierarchy. The understanding of the nonequilibrium properties of complex fluids such as the viscoelastic behavior of polymeric liquids, the rheological properties of Ferro fluids and liquid crystals subjected to magnetic fields, based on the architecture of their molecular constituents is useful to get a comprehensive view of the process. The analysis of simple physical particle models for complex fluids has developed from the molecular computation of basic systems (atoms, rigid molecules) to the simulation of macromolecular 'complex' system with a large number of internal degrees of freedom exposed to external forces [137, 138].

The most widely used simulation methods for molecular systems are Monte Carlo, Brownian dynamics and molecular dynamics. The microscopic approach represents the microstructural features of material by means of a large number of micromechanical elements (beads, platelet, rods) obeying stochastic differential equations. The evolution equations of the microelements arise from a balance of momentum at the elementary level. The Monte Carlo method is a stochastic strategy that relies on probabilities. The Monte Carlo sampling technique generates large numbers of configurations or microstates of equilibrated systems by stepping from one microstate to the next in a particular statistical ensemble. Random changes are made to the positions of the species present, together with their orientations and conformations where appropriate. Brownian dynamics are an efficient approach for simulations of large polymer molecules or colloidal particles in a small molecule solvent. Molecular dynamics is the most detailed molecular simulation method, which computes the motions of individual molecules. Molecular dynamics efficiently evaluates different configurational properties and dynamic quantities, which cannot generally be obtained by Monte Carlo [139, 140].

The first computer simulation of liquids was carried out in 1953. The model was an idealized two-dimensional representation of molecules as rigid disks. For macromolecular systems, the coarse-grained approach is widely used as the modeling process is simplified, hence becomes more efficient, and the characteristic topological features of the molecule can still be maintained. The level of detail for a coarse-grained model varies in different cases. The whole molecule can be represented by a single particle in a simulation and interactions between particles incorporate average properties of the whole molecule. With this approach, the number of degrees of freedom is greatly reduced [141].

On the other hand, a segment of a polymer molecule can also be represented by a particle (bead). The first coarse-grained model, called the "dumbbell" model (Fig. 1.5), was introduced in the 1930s. Molecules are treated as a pair of beads interacting via a harmonic potential. However, by using this model, it is possible to perform kinetic theory derivations and calculations for nonlinear rheological properties and solve some flow problems. The analytical results for the dumbbell models can also be used to check computer simulation procedures in molecular dynamics and Brownian dynamics [142, 143].

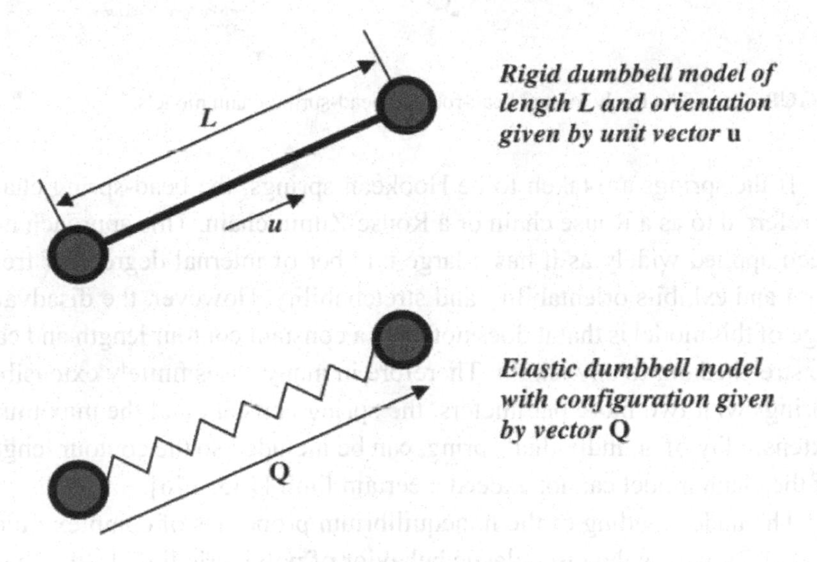
L
u
Rigid dumbbell model of length L and orientation given by unit vector u
Q
Elastic dumbbell model with configuration given by vector Q

FIGURE 1.5 The first coarse-grained models – the rigid and elastic dumbbell models.

The bead-rod and bead-spring model were introduced to model chain-like macromolecules. Beads in the bead-rod model (Fig. 1.6) do not represent the atoms of the polymer chain backbone, but some portion of the chain, normally 10 to 20 monomer units. These beads are connected by rigid and massless rods. While in the bead-spring model, a portion of the chain containing several hundreds of backbone atoms are replaced by a "spring" and the masses of the atoms are concentrated on the mass of beads [144].

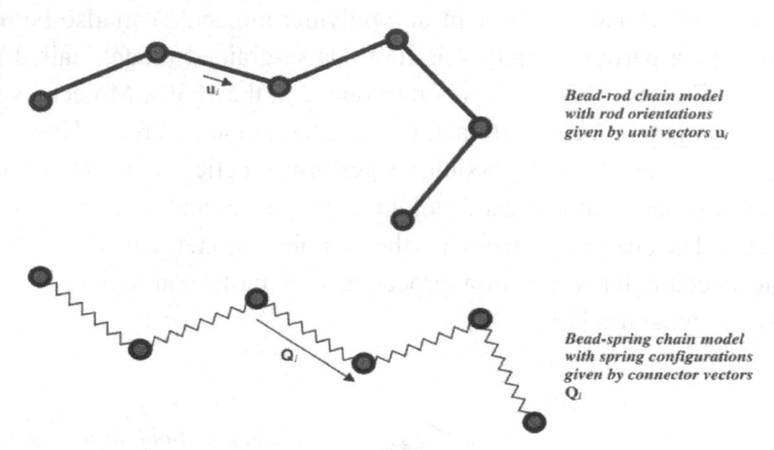

u_i
Bead-rod chain model
with rod orientations
given by unit vectors u_i
Q_i
Bead-spring chain model
with spring configurations
given by connector vectors
Q_i

FIGURE 1.6 The freely jointed bead-rod and bead-spring chain models.

If the springs are taken to be Hookean springs, the bead-spring chain is referred to as a Rouse chain or a Rouse-Zimm chain. This approach has been applied widely as it has a large number of internal degrees of freedom and exhibits orientability and stretchability. However, the disadvantage of this model is that it does not have a constant contour length and can be stretched out to any length. Therefore in many cases finitely extensible springs with two more parameters, the spring constant and the maximum extensibility of an individual spring, can be included so the contour length of the chain model cannot exceed a certain limit [145, 146].

The understanding of the nonequilibrium properties of complex fluids (Fig. 1.7) such as the viscoelastic behavior of polymeric liquids, the rheological properties of Ferro fluids and liquid crystals subjected to magnetic fields, based on the architecture of their molecular constituents [137].

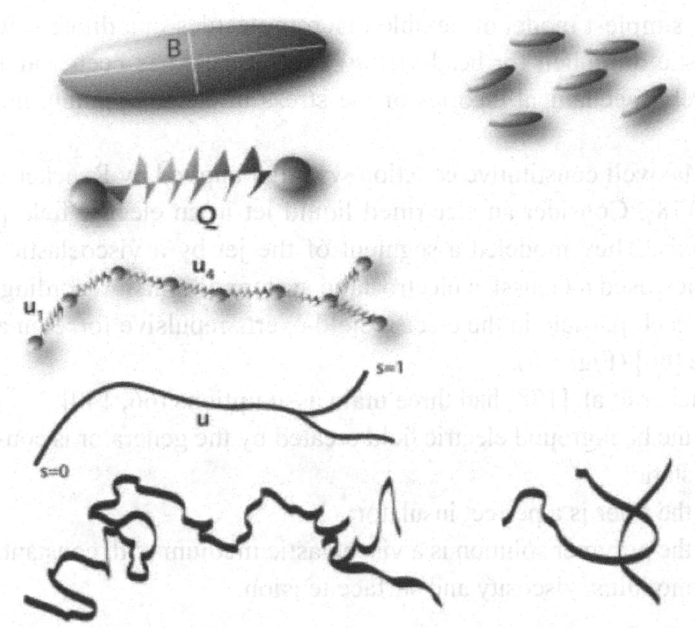

FIGURE 1.7 Simple microscopic models for complex fluids by using dumbbell model.

Dumbbell models are very crude representations of polymer molecules. Too crude to be of much interest to a polymer chemist, since it in no way accounts for the details of the molecular architecture. It certainly does not have enough internal degrees of freedom to describe the very rapid motions that contribute, for example, to the complex viscosity at high frequencies. On the other hand, the elastic dumbbell model is orientable and stretchable, and these two properties are essential for the qualitative description of steady-state rheological properties and those involving slow changes with time. For dumbbell models one can go through the entire program of endeavor—from molecular model for fluid dynamics—for illustrative purposes, in order to point the way towards the task that has ultimately to be performed for more realistic models. According to the researches, dumbbell models must, to some extend then, be regarded as mechanical playthings, somewhat disconnected from the real world of polymers. When used intelligently, however, they can be useful pedagogically and very helpful in developing a qualitative understanding of rheological phenomena [137, 147].

The simplest model of flexible macromolecules in a dilute solution is the elastic dumbbell (or bead-spring) model. This has been widely used for purely mechanical theories of the stress in electrospinning modeling [148].

A Maxwell constitutive equation was first applied by Reneker et al. in 2000 [178]. Consider an electrified liquid jet in an electric field parallel to its axis. They modeled a segment of the jet by a viscoelastic dumbbell. They used a Gaussian electrostatic system of units. According to this model, each particle in the electric field exerts repulsive force on another particle [66] (Fig. 1.8).

Reneker et al. [178] had three main assumptions [66, 149]:

a) the background electric field created by the generator is considered static;
b) the fiber is a perfect insulator;
c) the polymer solution is a viscoelastic medium with constant elastic modulus, viscosity and surface tension.

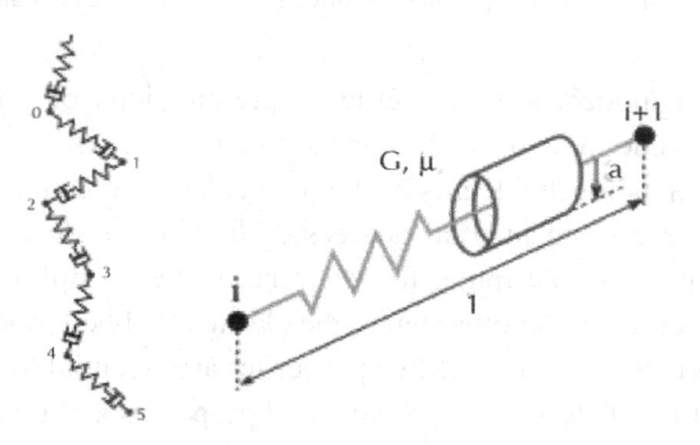

FIGURE 1.8 A schematic of one section of the model.

The researcher considered the governing equations for each bead as [149]:

$$\frac{d}{dt}\left(\pi a^2 l\right) = 0 \tag{90}$$

Therefore, the stress between these particles can be measured by [66]:

$$\frac{d\sigma}{dt}=G\frac{dl}{ldt}-\frac{G}{\eta}\sigma \tag{91}$$

The stress can be calculated by a Maxwell viscoelastic constitutive equation [150]:

$$\dot{\tau}=G\left(\varepsilon'-\frac{\tau}{\eta}\right) \tag{92}$$

where ε' is the Lagrangian axial strain [150]:

$$\varepsilon'\equiv\frac{\partial\dot{x}}{\partial\xi}.\hat{t}. \tag{93}$$

Equation of motion for beads can be written as [151]:

Mass × Acceleration = Viscous drag + Brownian motion force + force of one bead on another through the connector (94)

The momentum balance for a bead is [149]:

$$m\frac{d\upsilon}{dt}=\underbrace{-\frac{q^2}{l^2}}_{Coulomb\ \ forces}-\underbrace{qE}_{Electric\ \ force}+\underbrace{\pi a^2\sigma}_{Mechanical\ \ forces} \tag{95}$$

So the momentum conservation for model charges can be calculated as [152]:

$$m_i\frac{d\upsilon_i}{dt}=\underbrace{q_i\sum_{i\neq j}q_jK\frac{r_i-r_j}{\left|r_i-r_j\right|^3}}_{Coulomb\ \ forces}+\underbrace{q_iE}_{Electric\ \ force}\underbrace{+\pi a_{i,i+1}^2\sigma_{i,i+1}\frac{r_{i+1}-r_i}{\left|r_{i+1}-r_i\right|}-\pi a_{i-1,i}^2\sigma_{i-1,i}\frac{r_i-r_{i-1}}{\left|r_i-r_{i-1}\right|}}_{Mechanical\ \ forces} \tag{96}$$

Boundary condition assumptions: a small initial perturbation is added to the position of the first bead, the background electric field is axial and uniform and the first bead is described by a stationary equation. For solving these equations some dimensionless parameters are defined then by simplifying equations, the equations are solved by using boundary conditions [149, 152].

Now, an example for using this model for the polymer structure is mentioned. For a dumbbell consists of two (Fig. 1.9), which are connected with a nonlinear spring, the spring force law is given by [153]:

$$F = -\frac{HQ}{1 - Q^2/Q_0^2} \tag{97}$$

Now if we considered the model for the polymer matrix such as carbon nanotube, the rheological behavior can be obtained as [153, 154]:

$$\tau_{ij} = \tau_p + \tau_s \tag{98}$$

$$\tau_p = \underbrace{n_a \langle Q_a F_a \rangle}_{aggregated\ \ dumbbells} + \underbrace{n_f \langle Q_f F_f \rangle}_{free\ \ dumbbells} - nkT\delta_{ij} \tag{99}$$

$$\tau_s = \eta\dot{\gamma} \tag{100}$$

$$\lambda \langle Q.Q \rangle^{\nabla} = \delta_{ij} - \frac{c\langle Q.Q \rangle}{1 - tr\langle Q.Q \rangle / b_{max}} \tag{101}$$

The polymeric stress can be obtained from the following relation [153]:

$$\frac{\hat{\tau}_{ij}}{n_d kT} = \delta_{ij} - \frac{c\langle Q.Q \rangle}{1 - tr\langle Q.Q \rangle / b_{max}} \tag{102}$$

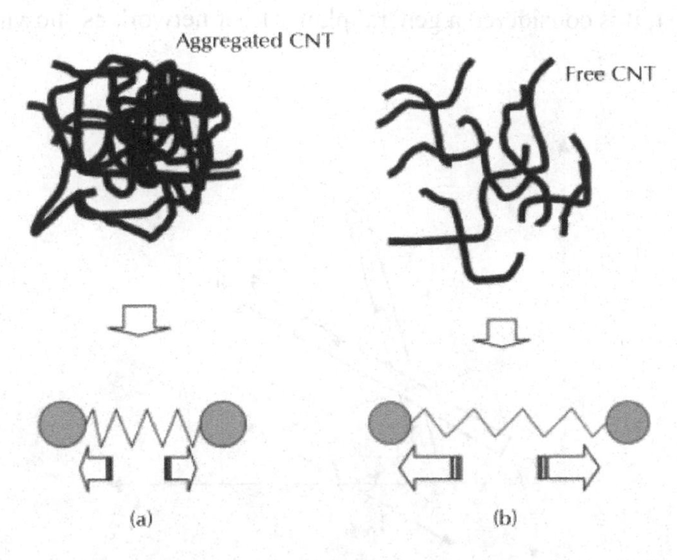

FIGURE 1.9 Modeling of two kinds of dumbbell sets. (a) aggregate FENE dumbbell, which has lower mobility and (b) free FENE dumbbell, which has higher mobility [153].

1.7.5 NETWORK MACROSCOPIC PROPERTIES

Some researchers analyze the process of phenomena in a network view. For example, in a thesis named "A 2D model for the electrospinning process," the researcher analyzes the electrospinning process with a new view of Feng solution. The external field was calculated on a 2D domain and interpolated back onto the jet coordinates [155]. On the other hand, some researches are done in analyzing the product properties of the process in a network view. Macroscopic properties of the multifunctional structure determine the final value of the any engineering product. The major objective in the determination of macroscopic properties is the link between atomic and continuum types of modeling and simulation approaches. The main advantage of the mesoscopic model is its higher computational efficiency than the molecular modeling without a loss of detailed properties at the molecular level. Peridynamic modeling of fibrous network is another promising method, which allows damage, fracture and long-range forces to be treated as natural components of the deformation of materials [156, 157].

At first, it is considered a general planar fiber network as shown in Fig. 1.10.

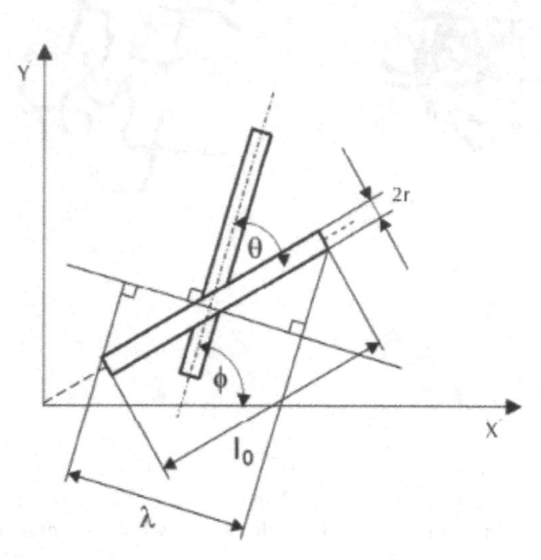

FIGURE 1.10 The fiber contact analysis [158].

The Poisson probability distribution can be applied to define the fiber segment length distribution for electrospun fabrics, a portion of the fiber between two neighboring contacts [158, 159]:

$$f(l) = \frac{1}{\bar{l}} \exp(-l/\bar{l}) \tag{103}$$

The total number fiber segments $\hat{N}$ in the rectangular region $b \times h$ can be obtained as:

$$\hat{N} = \{n.l_0(\langle\lambda\rangle + +2r) - 1\}.n.b.h \tag{104}$$

$$\langle\lambda\rangle = \int_0^{\varphi}\int_0^{\infty} \psi(\phi,l).\lambda(\phi).dl.d\phi \tag{105}$$

where, the dangled segments at fiber ends have been excluded.

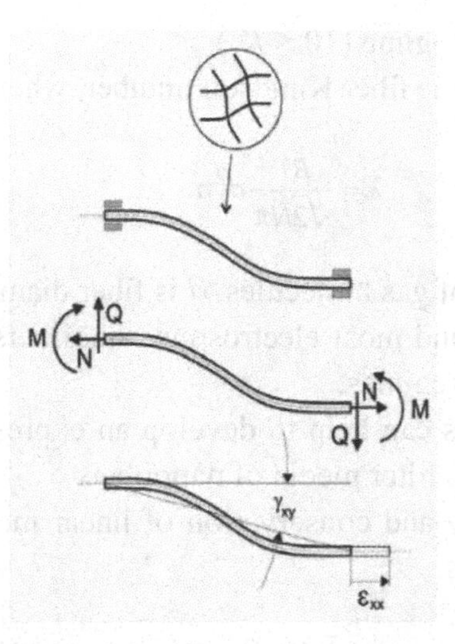

FIGURE 1.11 Fiber network 2D model.

The strain energy in fiber segments comes from bending, stretching and shearing modes of deformation can be calculated as [158]:

$$U = n.l_0.b.h\frac{1}{2}\iint\frac{E.A}{l}\varepsilon_{xx}^2.\psi(\phi,l).l.dl.d\phi + n.\{n.l_0.(\langle\lambda\rangle + 2r) - 1\}.b.h.\frac{1}{2} \tag{106}$$

$$\left\{\iint\frac{G.A}{l}.\gamma_{xy}^2.\psi(\phi,l).l.dl.d\phi + \iint\frac{3.E.I_m}{l^3}\gamma_{xy}^2.\psi(\phi,l).l.dl.d\phi\right\}$$

The strain energy fiber network (Fig. 1.11) for the representative volume element is equal to strain energy continuum element with effective material constant. The strain energy of the representative volume element under plane stress conditions are:

$$U = \frac{1}{2}.\langle\varepsilon_{ij}\rangle.C_{ijkl}.\langle\varepsilon_{kl}\rangle.\underbrace{V}_{b.h.2r} \tag{107}$$

Depending on the fiber diameter and the air thermal conditions, there are four different regimes of flow around a fiber:

a) Continuum regime ($K_N \leq 10^{-3}$),
b) Slip-flow regime ($10^{-3} \leq K_N \leq 0.25$),
c) Transient regime ($0.25 \leq K_N \leq 10$,

d) Free molecule regime ($10 \leq K_N$).

Here, $K_N = 2\lambda/d$ is the fiber Knudson number, where:

$$\lambda = \frac{RT}{\sqrt{2}N\pi} d^2 p \tag{108}$$

is the mean free path of gas molecules, d is fiber diameter, N is Avogadro number. Air flow around most electrospun nanofibers is typically in the slip or transition flow regimes.

This regime studies can help to develop an expression for predicting the permeability across filter media of nanofibers.

To solve continuity and conservation of linear momentum in the absence of inertial effects:

$$\nabla.\nu = 0 \tag{109}$$

$$\nabla p = \mu.\Delta\nu^2 \tag{110}$$

Permeability of a fibrous material is often presented as a function of fiber radius, r, and solid volume fraction α, of the medium. The continuum regime can be used for introducing an analytical expressions developed for 3D isotropic fibrous structures given as:

$$\frac{k}{r^2} = \frac{3r^2}{20\alpha}[-\ln(\alpha) - 0.931] \tag{111}$$

1.7.6 SCALING

The physical aspect of a phenomenon can use the language of differential equation, which represents the structure of the system by selecting the variables that characterize the state of it and certain mathematical constraint on the values of those variables can take on. These equations can predict the behavior of the system over a quantity such as time. For an instance, a set of continuous functions of time that describe the way the variables of the system developed over time starting from a given initial state [160]. In general, the renormalization group theory, scaling and fractal geometry, are applied to the understanding of the complex phenomena in physics, economics and medicine [161].

In more recent times, in statistical mechanics, the expression "scaling laws" has referred to the homogeneity of form of the thermodynamic and correlation functions near critical points, and to the resulting relations among the exponents that occur in those functions. From the viewpoint of scaling, electrospinning modeling can be studied in two ways, allometric and dimensionless analysis. Scaling and dimensional analysis actually started with Newton, and allometry exists everywhere in our daily life and scientific activity [161, 162].

1.7.6.1 ALLOMETRIC SCALING

Electrospinning applies electrically generated motion to spin fibers. So, it is difficult to predict the size of the produced fibers, which depends on the applied voltage in principal. Therefore the relationship between the radius of the jet and the axial distance from the nozzle is always the subject of investigation [163, 164]. It can be described as an allometric equation by using the values of the scaling exponent for the initial steady, instability and terminal stages [165].

The relationship between r and z can be expressed as an allometric equation of the form:

$$r \approx z^{b} \tag{112}$$

When the power exponent, $b = 1$ the relationship is isometric and when $b \neq 1$ the relationship is allometric [163, 166]. In another view, $b = -1/2$ is considered for the straight jet, $b = -1/4$ for instability jet and $b = 0$ for final stage [123, 164].

Due to high electrical force acting on the jet, it can be illustrated [163]:

$$\frac{d}{dz}\left(\frac{v^2}{2}\right) = \frac{2\sigma E}{\rho r} \tag{113}$$

Equations of mass and charge conservations applied here as mentioned earlier [163, 166–167].

From the above equations it can be seen that [112, 163]:

$$V \approx r^{-2},\ \sigma \approx r,\ E \approx r^{-2},\ \frac{d\upsilon^2}{dz} \approx r^{-2} \tag{114}$$

So it is obtained for the initial part of jet $r \approx z^{-1/2}$, $r \approx z^{-1/4}$ for the instable stage and $r \approx z^{0}$ for the final stage.

The charged jet can be considered as a one-dimensional flow as mentioned. If the conservation equations modified, they would change as [163]:

$$2\pi r \sigma^{\alpha} v + K\pi r^2 E = I \tag{115}$$

$$r \approx z^{-\alpha/(\alpha+1)} \tag{116}$$

where, α is a surface charge parameter, the value of α depends on the surface charge in the jet. When α = 0, no charge in jet surface, and in α = 1 use for full surface charge.

Allometric scaling equations are more widely investigated by different researchers. Some of the most important allometric relationships for electrospinning are presented in Table 1.2.

TABLE 1.2 Investigated scaling laws applied in electrospinning model.

Parameters	Equation	Ref.
The conductance and polymer concentration	$g \approx c^{\beta}$	[112]
The fiber diameters and the solution viscosity	$d \approx \eta^{\alpha}$	[164]
The mechanical strength and threshold voltage	$\overline{\sigma} \approx E_{threshold}^{-\alpha}$	[168]
The threshold voltage and the solution viscosity	$E_{threshold} \approx \eta^{1/4}$	[168]

TABLE 1.2 *(Continued)*

Parameters	Equation	Ref.
The viscosity and the oscillating frequency	$\eta \approx \omega^{-0.4}$	[168]
the volume flow rate and the current	$I \approx Q^{b}$	[167]
The current and the fiber radius	$I \approx r^{2}$	[169]
The surface charge density and the fiber radius	$\sigma \approx r^{3}$	[169]
The induction surface current and the fiber radius	$\phi \approx r^{2}$	[169]
The fiber radius and AC frequency	$r \approx \Omega^{1/4}$	[123]

β, α and b = scaling exponent.

1.7.6.2 DIMENSIONLESS ANALYSIS

One of the simplest, yet most powerful, tools in the physics is dimensional analysis in which there are two kinds of quantities: dimensionless and dimensional. Dimensionless quantities, which are without associated physical dimensions, are widely used in mathematics, physics, engineering, economics, and in everyday life (such as in counting). Numerous well-known quantities, such as π, e, and φ, are dimensionless. They are "pure" numbers, and as such always have a dimension of 1 [170–171].

Dimensionless quantities are often defined as products or ratios of quantities that are not dimensionless, but whose dimensions cancel out when their powers are multiplied [172].

In nondimensional scaling, there are two key steps:

(a) Identify a set of physically relevant dimensionless groups, and

(b) Determine the scaling exponent for each one.

Dimensional analysis will help you with step (a), but it cannot be applicable possibly for step (b).

A good approach to systematically getting to grips with such problems is through the tools of dimensional analysis (Bridgman, 1963). The dominant balance of forces controlling the dynamics of any process depends on the relative magnitudes of each underlying physical effect entering the set of governing equations [173]. Now, the most general characteristics parameters, which used in dimensionless analysis in electrospinning are introduced in Table 1.3.

TABLE 1.3 Characteristics parameters employed and their definitions.

Parameter	Definition
Length	R_0
Velocity	$\upsilon_0 = \dfrac{Q}{\pi R_0^2 K}$
Electric field	$E_0 = \dfrac{I}{\pi R_0^2 K}$
Surface charge density	$\sigma_0 = \bar{\varepsilon} E_0$
Viscose stress	$\tau_0 = \dfrac{\eta_0 \upsilon_0}{R_0}$

For achievement of a simplified form of equations and reduction a number of unknown variables, the parameters should be subdivided into characteristic scales in order to become dimensionless. Electrospinning dimensionless groups are shown in Table 1.4 [174].

TABLE 1.4 Dimensionless groups employed and their definitions.

Name	Definition	Field of application
Froude number	$Fr = \dfrac{\upsilon_0^2}{g R_0}$	The ratio of inertial to gravitational forces

TABLE 1.4 *(Continued)*

Name	Definition	Field of application
Reynolds number	$Re = \frac{\rho \upsilon_0 R_0}{\eta_0}$	The ratio of the inertia forces of the viscous forces
Weber number	$We = \frac{\rho \upsilon_0^2 R_0}{\gamma}$	The ratio of the surface tension forces to the inertia forces
Deborah number	$De = \frac{\lambda \upsilon_0}{R_0}$	The ratio of the fluid relaxation time to the instability growth time
Electric Peclet number	$Pe = \frac{2\bar{\varepsilon} \upsilon_0}{K R_0}$	The ratio of the characteristic time for flow to that for electrical conduction
Euler number	$Eu = \frac{\varepsilon_0 E^2}{\rho \upsilon_0^2}$	The ratio of electrostatic forces to inertia forces
Capillary number	$Ca = \frac{\eta \upsilon_0}{\gamma}$	The ratio of inertia forces of viscous forces
Ohnesorge number	$oh = \frac{\eta}{(\rho \gamma R_0)^{1/2}}$	The ratio of viscous force to surface force
Viscosity ratio	$r_\eta = \frac{\eta_p}{\eta_0}$	The ratio of the polymer viscosity to total viscosity
Aspect ratio	$\chi = \frac{L}{R_0}$	The ratio of the length of the primary radius of jet
Electrostatic force parameter	$\varepsilon = \frac{\bar{\varepsilon} E_0^2}{\rho \upsilon_0^2}$	The relative importance of the electrostatic and hydrodynamic forces
Dielectric constant ratio	$\beta = \frac{\varepsilon}{\bar{\varepsilon}} - 1$	The ratio of the field without the dielectric to the net field with the dielectric

The governing and constitutive equations can be transformed into a dimensionless form using the dimensionless parameters and groups.

1.7.7 SOME OF ELECTROSPINNING MODELS

The most important mathematical models for electrospinning process are classified in the Table 1.5. According to the year, advantages and disadvantages of the models:

TABLE 1.5 The most important mathematical models for electrospinning.

Researchers	Model	Year	Ref.
Taylor, G. I. Melcher, J. R.	Leaky dielectric model ☐ Dielectric fluid ☐ Bulk charge in the fluid jet considered to be zero ☐ Only axial motion ☐ Steady state part of jet	1969	[175]
Ramos	Slender body ☐ Incompressible and axi-symmetric and viscous jet under gravity force ☐ No electrical force ☐ Jet radius decreases near zero ☐ Velocity and pressure of jet only change during axial direction ☐ Mass and volume control equations and Taylor expansion were applied to predict jet radius	1996	[176]
Saville, D. A.	Electrohydrodynamic model ☐ The hydrodynamic equations of dielectric model was modified ☐ Using dielectric assumption ☐ This model can predict drop formation ☐ Considering jet as a cylinder (ignoring the diameter reduction) ☐ Only for steady state part of the jet	1997	[175]

TABLE 1.5 *(Continued)*

Researchers	Model	Year	Ref.
Spivak, A. Dzenis, Y.	Spivak and Dzenis model ▯ The motion of a viscose fluid jet with lower conductivity were surveyed in an external electric field ▯ Single Newtonian Fluid jet ▯ The electric field assumed to be uniform and constant, unaffected by the charges carried by the jet ▯ Use asymptotic approximation was applied in a long distance from the nozzle ▯ Tangential electric force assumed to be zero ▯ Using nonlinear rheological constitutive equation (Ostwald-de waele law), nonlinear behavior of fluid jet were investigated	1998	[101]
Jong Wook	Droplet formation model ▯ Droplet formation of charged fluid jet was studied in this model ▯ The ratio of mass, energy and electric charge transition are the most important parameters on droplet formation ▯ Deformation and break-up of droplets were investigated too ▯ Newtonian and Non-Newtonian fluids ▯ Only for high conductivity and viscous fluids	2000	[177]

TABLE 1.5 *(Continued)*

Researchers	Model	Year	Ref.
Reneker, D. H. Yarin, A. L.	Reneker model. ⍰ For description of instabilities in viscoelastic jets. ⍰ Using molecular chain theory, behavior of polymer chain of spring-bead model in electric field was studied. ⍰ Electric force based on electric field cause instability of fluid jet while repulsion force between surface charges make perturbation and bending instability. ⍰ The motion paths of these two cases were studied. ⍰ Governing equations: momentum balance, motion equations for each bead, Maxwell tension and columbic equations.	2000	[178]
Hohman, M. Shin, M.	Stability theory ⍰ This model is based on a dielectric model with some modification for Newtonian fluids. ⍰ This model can describe whipping, bending and Rayleigh instabilities and introduced new ballooning instability. ⍰ Four motion regions were introduced: dipping mode, spindle mode, oscillating mode, precession mode. ⍰ Surface charge density introduced as the most effective parameter on instability formation. ⍰ Effect of fluid conductivity and viscosity on nanofibers diameter were discussed. ⍰ Steady solutions may be obtained only if the surface charge density at the nozzle is set to zero or a very low value	2001	[60]

TABLE 1.5 *(Continued)*

Researchers	Model	Year	Ref.
Feng, J. J	Modifying Hohman model ⍰ For both Newtonian and nonNewtonian fluids ⍰ Unlike Hohman model, the initial surface charge density was not zero, so the "ballooning instability" did not accrue. ⍰ Only for steady state part of the jet ⍰ Simplifying the electric field equation, which Hohman used in order to eliminate Ballooning instability.	2002	[104]
Wan-Guo-Pan	Wan-Guo-Pan model ⍰ They introduced thermo-electro-hydro dynamics model in electrospinning process ⍰ This model is a modification on Spivak model which mentioned before ⍰ The governing equations in this model: Modified Maxwell equation, Navier-Stocks equations, and several rheological constitutive equation	2004	[126]
Ji-Haun	AC-electrospinning model ⍰ Whipping instability in this model was distinguished as the most effective parameter on uncontrollable deposition of nanofibers ⍰ Applying AC current can reduce this instability so make oriented nanofibers ⍰ This model found a relationship between axial distance from nozzle and jet diameter ⍰ This model also connected AC frequency and jet diameter	2005	[123]

TABLE 1.5 *(Continued)*

Researchers	Model	Year	Ref.
Roozemond (Eindhoven University and Technology)	Combination of slender body and dielectric model • In this model, a new model for viscoelastic jets in electrospinning were presented by combining these two models • All variables were assumed uniform in cross section of the jet but they changed in during z direction • Nanofiber diameter can be predicted	2007	[179]
Wan	Electromagnetic model • Results indicated that the electromagnetic field which made because of electrical field in charged polymeric jet is the most important reason of helix motion of jet during the process	2012	[180]
Dasri	Dasri model • This model was presented for description of unstable behavior of fluid jet during electrospinning • This instability causes random deposition of nanofiber on surface of the collector • This model described dynamic behavior of fluid by combining assumption of Reneker and Spivak models	2012	[181]

The most frequent numeric mathematical methods, which were used in different models, are listed in Table 1.6.

TABLE 1.6 Applied numerical methods for electrospinning.

Method	Ref.
Relaxation method	[104, 110, 182]
Boundary integral method (boundary element method)	[150, 177]
Semi-inverse method	[110, 123]
(Integral) control-volume formulation	[176]
Finite element method	[175]
Kutta-Merson method	[183]
Lattice Boltzmann method with finite difference method	[184]

1.8 ELECTROSPINNING SIMULATION

Electrospun polymer nanofibers demonstrate outstanding mechanical and thermodynamic properties as compared to macroscopic-scale structures. These features arc attributed to nanofiber microstructure [185, 186]. Theoretical modeling predicts the nanostructure formations during electrospinning. This prediction could be verified by various experimental condition and analysis methods, which called simulation. Numerical simulations can be compared with experimental observations as the last evidence [100, 187].

Parametric analysis and accounting complex geometries in simulation of electrospinning are extremely difficult due to the nonlinearity nature in the problem. Therefore, a lot of researches have done to develop an existing electrospinning simulation of viscoelastic liquids [182].

1.9 ELECTROSPINNING SIMULATION EXAMPLE

In order to survey of electrospinning modeling application, the main equations were applied for simulating the process according to the constants, which summarized in Table 1.7.

Mass and charge conservations allow ν and σ to be expressed in terms of R and E, and the momentum and E-field equations can be recast into two second-order ordinary differential equations for R and E. The slender-body theory (the straight part of the jet) was assumed to investigate the jet behavior during the spinning distance. The slope of the jet surface (R') is maximum at the origin of the nozzle. The same assumption has been used in most previous models concerning jets or drops. The initial and boundary conditions, which govern the process, are introduced as:

Initial values ($z = 0$):

$$\begin{aligned} &R(0) = 1 \\ &E(0) = E_0 \\ &\tau_{prr} = 2r_\eta \frac{R_0'}{R_0^3} \\ &\tau_{pzz} = -2\tau_{prr} \end{aligned} \tag{117}$$

Feng [104] indicated that $E(0)$ effect is limited to a tiny layer below the nozzle, which its thickness is a few percent of R_0. It was assumed that the shear inside the nozzle is effective in stretching of polymer molecules as compared with the following elongation.

Boundary values (z = c):

$$\begin{aligned} &R(\chi) + 4\chi R'(\chi) = 0 \\ &E(\chi) = E_\chi \\ &\tau_{prr} = 2r_\eta \frac{R_\chi'}{R_\chi^3} \\ &\tau_{pzz} = -2\tau_{prr} \end{aligned} \tag{118}$$

The asymptotic scaling can be stated as [104]:

$$R(z) \propto z^{-1/4} \tag{119}$$

Just above the deposit point ($z = \chi$), asymptotic thinning conditions applied. R drops towards zero and E approaches $E_¥$. The electric field is not equal to $E_¥$, so we assumed a slightly larger value, E_χ.

Table 1.7. Constants, which were used in electrospinning simulation.

Constant	Quantity
Re	$2.5.\ 10^{-3}$
We	0.1
Fr	0.1
Pe	0.1
De	10
E	1
β	40
χ	20
E_0	0.7
E_χ	0.5
r_η	0.9

The momentum, electric field and stress equations could be rewritten into a set of four coupled first order ordinary differential equations (ODE's) with the above-mentioned boundary conditions. The numerical relaxation method is chosen to solve the generated boundary value problem.

The results of these systems of equations are presented in Figs. 1.4 and 1.5, which matched quite well with the other studies that have been published [60, 78, 104, 105, 110].

The variation prediction of R, R,' ER^2, ER^{2}' and E versus axial position (z) are shown in Fig. 1.4. Physically, the amount of conductible charges reduces with decreasing jet radius. Therefore, to maintain the same jet current, more surface charges should be carried by the convection. Moreover, in the considered simulation region, the density of surface charge gradually increased. As the jet gets thinner and faster, electric conduction gradually transfers to convection. The electric field is mainly induced by the axial gradient of surface charge, thus it is insensitive to the thinning of the electrospun jet:

$$\frac{d(\sigma R)}{dz} \approx -\left(2R\frac{dR}{dz}\right) / Pe \tag{120}$$

Therefore, the variation of E versus z can be written as:

$$\frac{d(E)}{dz} \approx \ln\chi\left(\frac{d^2R^2}{dz^2}\right)/Pe \tag{121}$$

Downstream of the origin, E shoots up to a peak and then relaxes due to the decrease of electrostatic pulling force in consequence of the reduction of surface charge density, if the current was held at a constant value. However, in reality, the increase of the strength of the electric field also increases the jet current, which is relatively linear [78, 104]. As the jet becomes thinner downstream, the increase of jet speed reduces the surface charge density and thus E, so the electric force exerted on the jet and thus R' become smaller. The rates of R and R' are maximum at $z = 0$, and then relaxes smoothly downstream toward zero [104, 110]. According to the relation between R, E and z, ER^2 and $ER^{2'}$ vary in accord with parts (c) and (d) in Fig. 1.12.

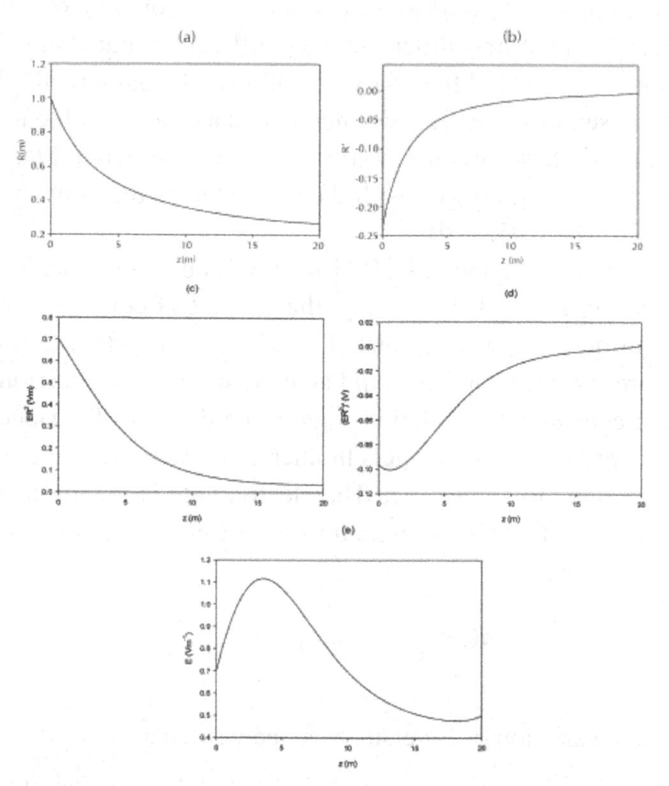

FIGURE 1.12 Solutions given by the electrospinning model for (a) R; (b) R'; (c) ER^2; (d) $ER^{2'}$ and (e) E.

Figure 1.13 shows the changes of axial, radial shear stress and the difference between them, the tensile force (*T*) versus *z*. The polymer tensile force is much larger in viscoelastic polymers because of the strain hardening. T also has an initial rise, because the effect of strain hardening is so strong that it overcomes the shrinking radius of the jet. After the maximum value of T, it reduces during the jet thinning. As expected, the axial polymer stress rises, because the fiber is stretched in axial direction, and the radial polymer stress declines. The variation of T along the jet can be nonmonotonic, however, meaning the viscous normal stress may promote or resist stretching in a different part of the jet and under different conditions [104, 110].

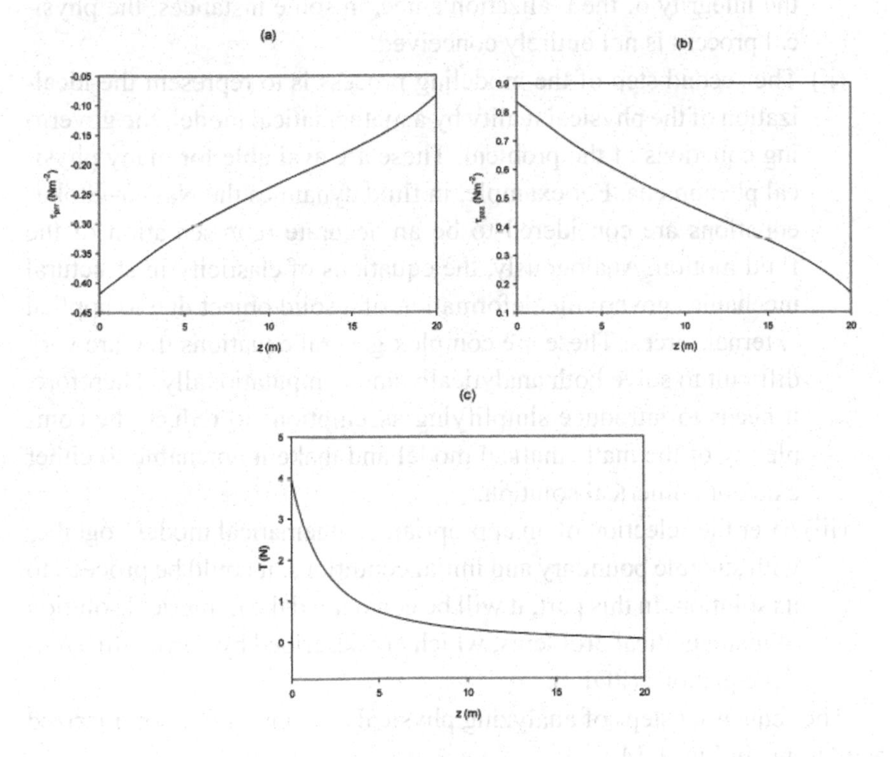

FIGURE 1.13 Solutions given by the electrospinning model for (a) τ_{prr}; (b) τ_{pzz} and (c) T.

1.10 APPLIED NUMERICAL METHODS FOR ELECTROSPINNING

Mathematics is indeed the language of science and the more proficient one is in the language the better. There are three main steps in the computational modeling of any physical process: (i) problem definition, (ii) mathematical model, and (iii) computer simulation [188–190].

(i) The first natural step is to define an idealization of our problem of interest in terms of a set of affiliate quantities, which it would be wanted to measure. In defining this idealization we expect to obtain a well-posed problem, this is one that has a unique solution for a given set of parameters. It might not always be possible to insure the integrity of the realization since, in some instances, the physical process is not entirely conceived.

(ii) The second step of the modeling process is to represent the idealization of the physical reality by a mathematical model: the governing equations of the problem. These are available for many physical phenomena. For example, in fluid dynamics the Navier–Stokes equations are considered to be an accurate representation of the fluid motion. Analogously, the equations of elasticity in structural mechanics govern the deformation of a solid object due to applied external forces. These are complex general equations that are very difficult to solve both analytically and computationally. Therefore, it needs to introduce simplifying assumptions to reduce the complexity of the mathematical model and make it amenable to either exact or numerical solution.

(iii) After the selection of an appropriate mathematical model, together with suitable boundary and initial conditions, it could be proceed to its solution. In this part, it will be considered the numerical solution of mathematical problems, which are described by partial differential equations (PDEs).

The sequential steps of analyzing physical systems can be summarized as presents in Fig. 1.14.

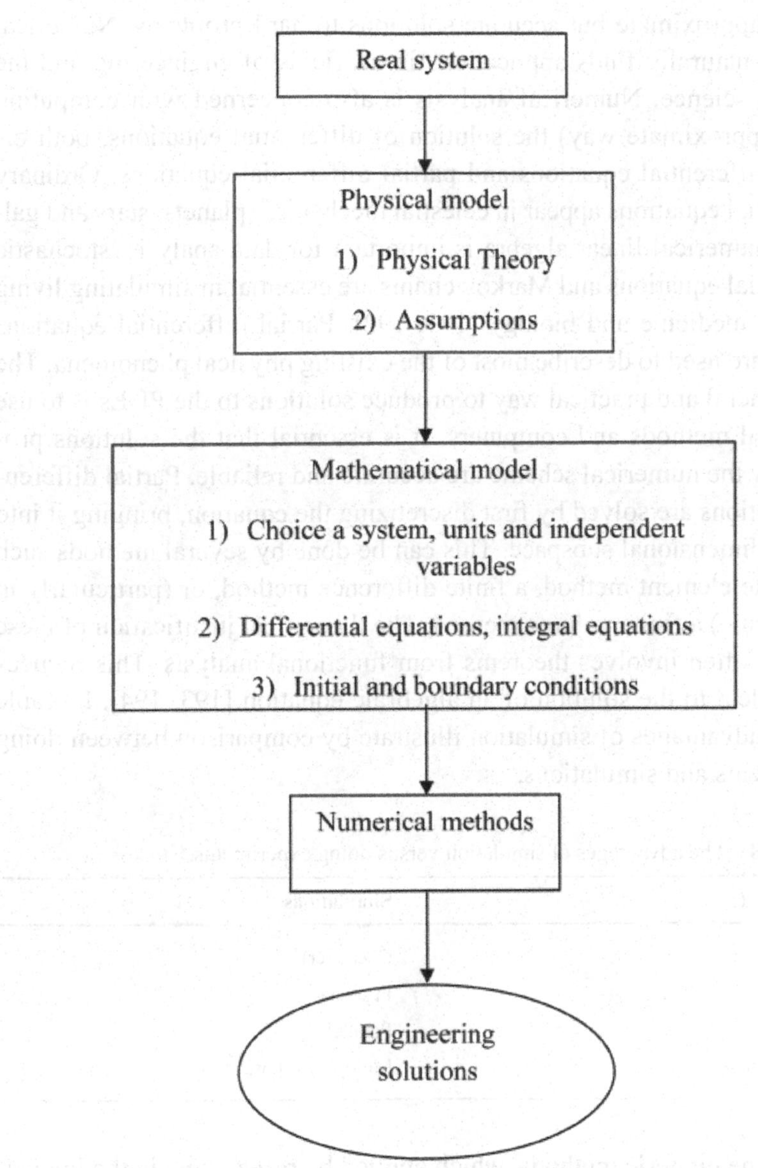

FIGURE 1.14 The diagram of sequential steps of analyzing physical systems.

Numerical analysis is the study of algorithms, which uses numerical approximations for the problems of mathematical analysis. The overall goal of the field of numerical analysis is the design and analysis of techniques to give approximate but accurate solutions to hard problems. Numerical analysis naturally finds applications in all fields of engineering and the physical science. Numerical analysis is also concerned with computing (in an approximate way) the solution of differential equations, both ordinary differential equations and partial differential equations. Ordinary differential equations appear in celestial mechanics (planets, stars and galaxies), numerical linear algebra is important for data analysis, stochastic differential equations and Markov chains are essential in simulating living cells for medicine and biology [191, 192]. Partial differential equations (PDEs) are used to describe most of the existing physical phenomena. The most general and practical way to produce solutions to the PDEs is to use numerical methods and computers. It is essential that the solutions produced by the numerical scheme are accurate and reliable. Partial differential equations are solved by first discretizing the equation, bringing it into a finite-dimensional subspace. This can be done by several methods such as a finite element method, a finite difference method, or (particularly in engineering) a finite volume method. The theoretical justification of these methods often involves theorems from functional analysis. This reduces the problem to the solution of an algebraic equation [193, 194]. In Table 1.8, the advantages of simulation illustrate by comparison between doing experiments and simulations.

TABLE 1.8 The advantages of simulation versus doing experiments.

Experiments	Simulations
Expensive	Cheap (er)
Slow	Fast (er)
Sequential	Parallel
Single-purpose	Multiple-purpose

Now, some numeric methods, which applied by researchers in the investigation on electrospinning models and simulations, are over looked.

1.10.1 KUTTA-MERSON METHOD

One hundred years ago, C. Runge [195, 196] was completing his famous paper. This work, published in 1895, extended the approximation method of Euler to a more elaborate scheme, which was able to have a greater accuracy by using Taylor series expansion. The idea of Euler was to bring up the solution of an initial value problem forward by a sequence of small time-steps. In each step, the rate of change of the solution is treated as constant and is found from the formula for the derivative evaluated at the beginning of the step.

The system of ordinary differential equations (ODEs) arising from the application of a spatial discretization of a system of PDEs can be very large, especially in three-dimensional simulations. Consequently, the constraints on the methods used for integrating these systems are somewhat different from those, which have taken much of the advance of numerical methods for initial value problems. On their high accuracy and modest memory requirements, the Runge–Kutta methods have become popular for simulations of physical phenomena. The classical fourth-order Runge–Kutta method requires three memory locations per dependent variable but low-storage methods requiring only two memory locations per dependent variable can be derived. This feature is easily achieved by a third-order Runge–Kutta method but an additional stage is required for a fourth-order method. Since the primary cost of the integration is in the evaluation of the derivative function, and each stage requires a function evaluation, the additional stage represents a significant increase in expense. For the same reason, error checking is generally not performed when solving very large systems of ODEs arising from the discretization of PDEs [196–198].

In 2004 and in 2006, Reznik and et al. [199, 200] experimentally and theoretically studied on the shape evolution of small compound droplets in the normal form and at the exit of a core shell system in the presence of a sufficiently strong electric field, respectively.

In the first study they considered an axisymmetric droplet of an incompressible conducting viscous liquid on an infinite conducting plate. It was neglected in the gravity effects so the stationary shape of the droplet is spherical. The droplet shaped (Fig. 1.15) as a spherical segment rests on the plate with a static contact angle of the liquid/air/solid system.

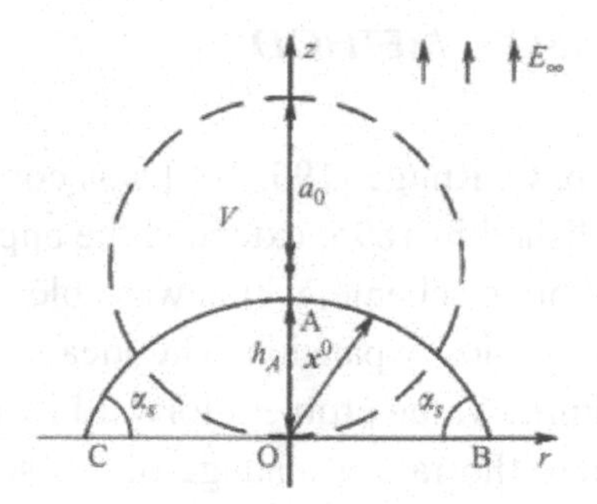

FIGURE 1.15 The initial shape of the droplet at the moment when the electric field is to be applied.

They estimated the relevant characteristic dimensionless parameters of the problem and used the Stokes equations for the liquid motion within the droplet then the equations solved numerically using the Kutta–Merson method [200].

In the second work, a core shell nozzle (Fig. 1.16.) consists of a central cylindrical pipe and a concentric annular pipe surrounding it. When the process took place without an applied electric field, the outer surface of the droplet and the interface between its components acquire near-spherical equilibrium shapes owing to the action of the surface and interfacial tension, respectively. If after the establishment of the equilibrium shape an electric field is applied to the compound nozzle and droplet attached to an electrode immersed in it, with a counter electrode, say a metal plate, located at some distance from the droplet tip, the latter undergoes stretching under the action of Maxwell stresses of electrical origin. At first, the problem was defined then the droplet evolution determined by time stepping using the Kutta-Merson method.

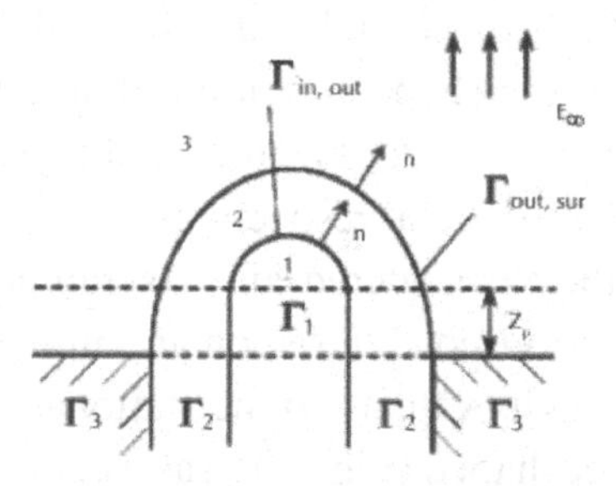

FIGURE 1.16 Core shell droplet at the exit of a core shell nozzle.

Both the core and the shell fluids are considered as leaky dielectrics, whose electric relative permittivities and conductivities are denoted. The electric boundary conditions describe the jump in the normal component of the electric field and electric induction at the boundaries. The governing equations rendered dimensionless and subject to the boundary conditions were solved with the aid of equivalent boundary integral formulations in which a set of corresponding integral equations was solved [199].

In 2010, Holzmeister et al. [183] considered a simple two-dimensional model can be used for describing the formation of barb electrospun polymer nanowires with a perturbed swollen cross-section and the electric charges "frozen" into the jet surface. This model is integrated numerically using the Kutta-Merson method with the adoptable time step. The result of this modeling is explained theoretically as a result of relatively slow charge relaxation compared to the development of the secondary electrically driven instabilities, which deform jet surface locally. When the disparity of the slow charge relaxation compared to the rate of growth of the secondary electrically driven instabilities becomes even more pronounced, the barbs transform in full-scale long branches. The competition between charge relaxation and rate of growth of capillary and electrically driven secondary localized perturbations of the jet surface is affected not only by the electric conductivity of polymer solutions but also by their viscoelasticity. Moreover, a nonlinear theoretical model was able to resemble the main morphological trends recorded in the experiments.

1.10.2 FINITE ELEMENT METHOD

The finite element method is a numerical analysis technique for obtaining approximate solutions to a vast variety of engineering problems. All finite element methods involve dividing the physical systems, such as structures, solid or fluid continua, into small subregions or elements. Each element is an essentially simple unit, the behavior of which can be readily analyzed. The complexities of the overall systems are accommodated by using large numbers of elements, rather than by resorting to the sophisticated mathematics required by many analytical solutions [201, 202].

A typical finite element analysis on a software system requires the following information [203]:

1. Nodal point spatial locations (geometry)
2. Elements connecting the nodal points
3. Mass properties
4. Boundary conditions or restraints
5. Loading or forcing function details
6. Analysis options

And the FEM Solution Process [204]:

1. Divide structure into pieces (elements with nodes) (discretization/meshing)
2. Connect (assemble) the elements at the nodes to form an approximate system of equations for the whole structure (forming element matrices)
3. Solve the system of equations involving unknown quantities at the nodes (e.g., displacements)
4. Calculate desired quantities (e.g., strains and stresses) at selected elements

One of the main attractions of finite element method is the ease with which they can be applied to problems involving geometrically complicated systems. The price that must be paid for flexibility and simplicity of individual elements is in the amount of numerical computation required. Very large sets of simultaneous algebraic equations have to be solved, and this can only be done economically with the aid of digital computers [205]. Therefore, the advantage of this method is that for a smooth problem where the derivatives of the solution are well behaved, the computational cost increases algebraically while the error decreases exponentially fast and the disadvantage of it is that the method leads to nonsingular systems of equations that can easily solve by standard methods of solution. This is not the case for time-dependent problems where numerical errors may grow unbounded for some discretization [189].

The finite element method is used for spatial discretization. Special numerical methods are used and developed for viscoelastic fluid flow, including the DEVSS and log-conformation techniques. For describing the moving sharp interface between the rigid particle and the fluid both ALE (Arbitrary Lagrangian Euler) and XFEM (extended finite element) techniques are being developed and employed [206]. In the development of numerical algorithms for the stable and accurate solution of viscoelastic flow problems, like electrospinning process, applying the finite element method to solve constitutive equations of the differential type is useful [207].

In 2011 Chitral et al. [208] analyzed the electrospinning process based on an existing electrospinning model for viscoelastic liquids using a finite element method. Four steady-state equations concluding Coulomb force, an electric force imposed by the external electric field, a viscoelastic force, a surface tension force, a gravitational force, and an air drag force were solved as a set of equations with this method.

1.10.3 BOUNDARY INTEGRAL METHOD (BOUNDARY ELEMENT METHOD)

Boundary integral equations are a classic tool for the analysis of boundary value problems for partial differential equations. The term "boundary element method" (BEM) denotes any method for the approximate numerical solution of these boundary integral equations. The approximate solution of the boundary value problem obtained by BEM has the distinguishing feature that it is an exact solution of the differential equation in the domain and is parameterized by a finite set of parameters living on the boundary [209] (Fig. 1.17.).

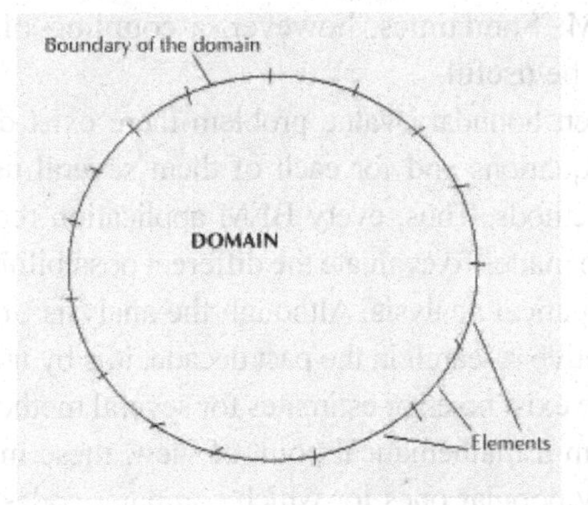

FIGURE 1.17 The idea of boundary element method.

The BEM has some advantages over other numerical methods like finite element methods (FEM) or finite differences [209–211]:

1. Only the boundary of the domain needs to be discretized. Especially in two dimensions where the boundary is just a curve this allows very simple data input and storage methods.
2. Exterior problems with unbounded domains but bounded boundaries are handled as easily as interior problems.
3. In some applications, the physically relevant data are given not by the solution in the interior of the domain but rather by the boundary values of the solution or its derivatives. These data can be obtained directly from the solution of boundary integral equations, whereas boundary values obtained from FEM solutions are in general not very accurate.
4. The solution in the interior of the domain is approximated with a rather high convergence rate and moreover, the same rate of convergence hold for all derivatives of any order of the solution in the domain. There are difficulties, however, if the solution has to be evaluated close to, but not on the boundary.

Some main difficulties with BEM are the following [209, 211, 212]:

1. Boundary integral equations require the explicit knowledge of a fundamental solution of the differential equation. This is available only for linear partial differential equations with constant or some specifically variable coefficients. Problems with inhomogeneities or nonlinear differential equations are in general not accessible by pure BEM. Sometimes, however, a coupling of FEM and BEM proves to be useful.
2. For a given boundary value problem there exist different boundary integral equations and for each of them several numerical approximation methods. Thus, every BEM application requires that several choices be made. To evaluate the different possibilities, one needs a lot of mathematical analysis. Although the analysis of BEM has been a field of active research in the past decade, it is by no means complete. Thus there exist no error estimates for several methods that are widely used. From a mathematical point of view, these methods, which include very popular ones for which computer codes are available, are in an experimental state, and there might exist problems of reliability.

3. The reason for the difficulty of the mathematical analysis is that boundary integral equations frequently are not ordinary Fredholm integral equations of the second kind. The classical theory of integral equations and their numerical solution concentrates on the second kind integral equations with regular kernel, however. Boundary integral equations may be of the first kind, and the kernels are in general singular. If the singularities are not integrable, one has to regularize the integrals, which are then defined in a distributional sense. The theoretical framework for such integral equations is the theory of pseudo-differential operators. This theory was developed 20 years ago and is now a classic part of Mathematical Analysis, but it is still not very popular within Applied Mathematics.
4. If the boundary is not smooth but has corners and edges, then the solution of the boundary value problem has singularities at the boundary. This happens also if the boundary conditions are discontinuous, for example, in mixed boundary value problems. BEM clearly has to treat these singularities more directly than FEM. Because the precise shape of the singularities frequently contains important information, e.g. stress intensity factors in fracture mechanics, this is a positive aspect of BEM. But besides practical problems with the numerical treatment of these singularities, nonsmooth domains also present theoretical difficulties. These have so far been satisfactorily resolved only for two-dimensional problems. The analysis of BEM for three-dimensional domains with corners and edges is still in a rather incomplete stage.

The paper of Kowalewski, which published in 2009, can be effectively addressed by use of a BEM. In essence, the boundary element method is a statement of the electrostatic problem (Poisson equation) in terms of boundary integrals, as such, it involves only the discretization of boundary surfaces, which in our case would be the electrode surfaces and the outer shell of the fiber. The model was, essentially, a time-dependent three-dimensional generalization of known slender models with the following differences:

(i) The electric field induced by the generator and by the charges on the fiber is explicitly resolved, instead of being approximated from local parameters

(ii) Electrical conductivity is neglected. Indeed, the convection of surface charges is believed to strongly overcomes bulk conduction at locations distant from the Taylor cone by a few fiber radii, since we are mostly interested in the description of the bending instability, this assumption appears reasonable [213].

1.10.4 (INTEGRAL) CONTROL-VOLUME FORMULATION

All the laws of mechanics are written for a system, which is defined as an arbitrary quantity of mass of fixed identity. Everything external to this system is denoted by the term surroundings, and the system is separated from its surroundings by its boundaries. The laws of mechanics then state what happens when there is an interaction between the system and its surroundings [214].

Typically, to understand how a given physical law applies to the system under consideration, one first begins by considering how it applies to a small, control volume, or "representative volume." There is nothing special about a particular control volume, it simply represents a small part of the system to which physical laws can be easily applied. This gives rise to what is termed a volumetric, or volume-wise formulation of the mathematical model [215, 216].

In fluid mechanics and thermodynamics, a control volume is a mathematical abstraction employed in the process of creating mathematical models of physical processes. In an inertial frame of reference, it is a volume fixed in space or moving with constant velocity through which the fluid (gas or liquid) flows. The surface enclosing the control volume is referred to as the control surface [217, 218].

Control-volume analysis is "more equal," being the single most valuable tool to the engineer for flow analysis. It gives "engineering" answers, sometimes gross and crude but always useful [214].

The advantage of this method, over that of the finite-element method, is the following. In the finite-element method, it is necessary to construct a grid over the flaw as well as the entire region surrounding the flaw, and to solve for the fields at all points on the grid. In contrast, in the volume-integral method, it is only necessary to construct a grid over the flaw and solve for the currents in the flaw; the Green's function takes care of all regions outside the flaw. This removes the complicated gridding requirements of the finite-element method, and reduces the size of the problem tremendously. That is the reason which this method can obtain much more accurate probe responses, and in much less time than a finite-element code, while running on a small personal computer or workstation. And without the complicated gridding, problems can be set up much more quickly and easily [219].

The famous researcher in electrospinning process, Feng, considered the steady stretching process is important in that it not only contributes to the thinning directly for Newtonian flows. The jet is governed by four steady-state equations representing the conservation of mass and electric charges, the linear momentum balance, and Coulomb's law for the electric field, which used control-volume balance for analyzing them [220].

1.10.5 RELAXATION METHOD

Relaxation method, an alternative to the Newton iteration method, is a method of solving simultaneous equations by guessing a solution and then reducing the errors that result by successive approximations until all the errors are less than some specified amount. Relaxation methods were developed for solving nonlinear systems and large sparse linear systems, which arose as finite-difference discretizations of differential equations [221, 222]. These iterative methods of relaxation should not be confused with "relaxations" in mathematical optimization, which approximate a difficult problem by a simpler problem, whose "relaxed" solution provides information about the solution of the original problem [223]. In solving PDEs problem with this method, it's necessary to turn them to the ODEs equations. Then ODEs have to be replaced by approximate finite difference equations. The relaxation method determines the solution by starting with a guess and improving it, iteratively. During the iterations the solution is improved and the result relaxes towards the true solution. Notice that the number of mesh points may be important for the properties of the numerical procedure. In relaxsetting increase the number of mesh points if the convergence towards steady state seems awkward [224, 225].

The advantage of this method is a relative freedom in its implementation. It can be used for smooth problem. This is useful in particular when an analytical solution to the model is not available and can handle models, which exhibit saddle-point stability. Therefore, a "large stepping" in the direction of the defect is possible, while the termination point is defined by the condition that the vector field becomes orthogonal. The relaxation algorithm can easily cope with a large number of problems, which arise frequently in the context of multidimensional, infinite-time horizon optimal control problems [226, 227].

A researcher in a thesis named "A Model for Electrospinning Viscoelastic Fluids" studied the electrospinning process of rewriting the momentum, electric field and stress equations as a set of six coupled first order ordinary differential equations. It made a boundary value problem, which can best be solved by a numerical relaxation method. Within relaxation methods the differential equations are replaced by finite difference equations on a certain mesh of points covering the range of integration. During iteration (relaxation) all the values on the mesh are adjusted to bring them into closer agreement with the finite difference equations and the boundary conditions [228].

1.10.6 LATTICE BOLTZMANN METHOD WITH FINITE DIFFERENCE METHOD

Lattice Boltzmann methods (LBM) (or Thermal Lattice Boltzmann methods (TLBM)) are a class of computational fluid dynamics (CFD) methods for fluid simulation. Instead of solving the Navier–Stokes equations, the discrete Boltzmann equation is solved to simulate the flow of a Newtonian fluid with collision models such as Bhatnagar-Gross-Krook (BGK). LBM is a relatively new simulation technique for complex fluid systems and has attracted interest from researchers in computational physics. Unlike the traditional CFD methods, which solve the conservation equations of macroscopic properties (i.e., mass, momentum, and energy) numerically, LBM models the fluid consisting of fictive particles, and such particles perform consecutive propagation and collision processes over a discrete lattice mesh. Due to its particulate nature and local dynamics, LBM has several advantages over other conventional CFD methods, especially in dealing with complex boundaries, incorporating of microscopic interactions, and parallelization of the algorithm. A different interpretation of the lattice Boltzmann equation is that of a discrete-velocity Boltzmann equation. By simulating streaming and collision processes across a limited number of particles, the intrinsic particle interactions evince a microcosm of viscous flow behavior applicable across the greater mass. The LBM is especially useful for modeling complicated boundary conditions and multi phase interfaces [229–231]. The idea that LBE is a discrete scheme of the continuous Boltzmann equation also provides a way to improve the computational efficiency and accuracy of LBM. From this idea, the discretization of the phase space and the configuration space can be done independently. Once the phase space is discretized, any standard numerical technique can serve the purpose of solving the discrete velocity Boltzmann equation. It is not surprising that the finite difference, finite volume, and finite element methods have been introduced into LBM in order to increase computational efficiency and accuracy by using nonuniform grids [232]. The lattice Boltzmann method has a more detailed microscopic description than a classical finite difference scheme because the LBM approach includes a minimal set of molecular velocities of the particles. In addition, important physical quantities, such as the stress tensor, or particle current, is directly obtained from the local information. However, the LBM scheme may require more memory than the corresponding finite difference scheme. Another motivation is that the boundary conditions are more or less naturally imposed for a given numerical scheme [233]. In the ordinary LBM, this discrete BGK equation is discretized into a special finite difference form in which the convection term does not include the numerical error. But considering the collision term, this scheme is of second-order accuracy. On the other hand, the discrete BGK equation can be discretized in some other finite-difference schemes or in the finite-volume method and other computational techniques for the partial differential equations, and these techniques are considered as a natural extension of numerical calculations [234–236].

There is a method, which couples lattice Boltzmann method for the fluid with a molecular dynamics model for polymer chain. The lattice Boltzmann equation using single relaxation time approximations (Fig. 1.18.).

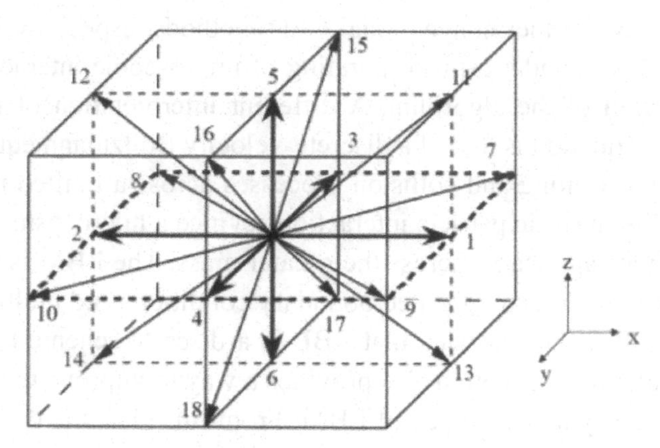

FIGURE 1.18 Lattice scheme.

In a thesis named "Modeling Electrospinning Process and a Numerical Scheme Using Lattice Boltzmann Method to Simulate Viscoelastic Fluid Flows," the researcher wrote the conservation of mass and momentum based on the lattice scheme which presented above. It was assumed that the fluid is incompressible and the continuum/macroscopic governing equations were written then forward step finite di*ff*erence method in time applied for solving obtained equations [237].

1.10.7 SEMI-INVERSE METHOD

From the mathematical point of view, for the problems which the analytical solution may be very hard to attain, even in the simplest boundary value problem, because the set of equation forms a nontrivial system of nonlinear, partial differential equations generating often non unique solutions. To solve the resulting boundary-value problems, inverse techniques can be used to provide simple solutions and to suggest experimental programs for the determination of response functions. Two powerful methods for inverse investigations are the so-called inverse method and the semi-inverse method. They have been used in elasticity theory as well as in all fields of the mechanics of continua [238–240].

In the inverse method, a solution is found in priori such that it satisfies the governing equation and boundary condition. We can obtain the solution through this way for a luck case, but it is not a logical way. For the semiinverse method, certain assumptions about the components of displacement strain are made at the beginning. Then, the solution is confined by satisfying the equations of equilibrium and the boundary conditions [240].

In the semiinverse method is systematically studied and many examples are given to show how to establish a variational formulation for a nonlinear equation. From the given examples, we found that it is difficult to find a variational principle for nonlinear evolution equations with nonlinear terms of any orders [241].

In the framework of the theory of Continuum Mechanics, exact solutions play a fundamental role for several reasons. They allow investigating in a direct way the physics of various constitutive models to understand in depth the qualitative characteristics of the differential equations under investigation and they provide benchmark solutions of complex problems.

The Mathematical method used to determine these solutions is usually called the semiinverse method. This is essentially a heuristic method that consists in formulating a priori a special ansatz on the geometric and/or kinematical fields of interest, and then introducing this ansatz into the field equations. Luck permitting, these field equations reduce to a simple set of equations and then some special boundary value problems may be solved. Another important aspect in the use of the semiinverse method is associated with fluid dynamics with the emergence of secondary flows and in solid mechanics with latent deformations. It is clear that "Navier-Stokes fluid" and an "isotropic incompressible hyperelastic material" are intellectual constructions [238, 242].

The most interesting features of this method are its extreme simplicity and concise forms of variational functionals for a wide range of nonlinear problems [243, 244]. An advantage of the semiinverse method is that it can provide a powerful mathematical tool in the search for variational formulations for a rather wide class of physic problems without using the well-known Lagrange multipliers, which can result in variational crisis (the constrains can not eliminated after the identification of the multiplier or the multipliers become zero), furthermore, to use the Lagrange multipliers, we must have a known variational principle at hand, a situation which not always occurs in continuum physics [245, 246]. Although, using this method has not any special difficulties, there are primary difficulties with ill-posed problems [247, 248].

In 2004, Wan and et al. [249] studied in a model of steady state jet, which introduced by Spivak but they considered the couple effects of thermal, electricity, and hydrodynamics. Therefore, the model consists of modified Maxwell's equations governing electrical field in a moving fluid, the modified Navier-Stokes equations governing heat and fluid flow under the influence of electric field, and constitutive equations describing behavior of the fluid. The set of conservation laws can constitute a closed system when it is supplemented by appropriate constitutive equations for the field variables such as polarization. The most general theory of constitutive equations determining the polarization, electric conduction current, heat flux, and Cauchy stress tensor has been developed by Eringen and Maugin. By the semiinverse method, it could be obtained various variational principles for electrospinning. After obtaining a set of equations, they were solved by the semi – inverse method.

1.11 CONCLUSION

Comprehensive investigation on the nanoscience and the related technologies is an important topic these days. Due to the rising interest in nanoscale material properties, studying on the effective methods of producing nanomaterial such as the electrospinning process plays an important role in the new technologies progress. Since the electrospinning process is dependent on a lot of different parameters, changing them will lead to significant variations in the process. In this paper, we have attempted to analyze the each part of process in detail and investigate some of the most important relationships from ongoing research into the fundamental physics that govern the electrospinning process. The idea has been to provide a few guiding principles for those who would use electrospinning to fabricate materials with the initial scrutiny. We have outlined the flow procedure involved in the electric field and those associated with the jetting process. The relevant processes are the steady thinning jet, whose behavior can be understood quantitatively using continuum equations of electrohydrodynamics, and the ensuing fluid dynamical instabilities that give rise to whipping of the jet. The current carried by the jet is of critical importance to the process and some scaling aspects concerning the total measured current have been discussed. The basics of electrospinning modeling involve mass conservation, electric charge conservation, momentum balance, coulomb's law and constitutive equations (Ostwald-de Waele power law, Giesekus equation and Maxwell equation), which are discussed in detail. These relations play an important role in setting final features. The dominant balance of forces controlling the dynamics of the electrospinning process depends on the relative magnitudes of each underlying physical effect entering the set of governing equations. Due to the application of the obtained nano web and material for different usages, surveying on microscopic and macroscopic properties of the product is necessary. Therefore, for achieving to this purpose, the relationships are pursued. Also for providing better understanding of electrospinning simulation, a model has been implemented for a straight jet in a relaxation method and simulated for the special viscoelastic polymers. The model capability to predict the behavior of the process parameters was demonstrated using simulation in the last part. The plots obviously showed the changes of each parameter versus axial position. During the jet thinning, the electric field shoots up to a peak and then relaxes. The tensile force also has an initial rise and then reduces. All the plots show similar behavior with the results of other researchers. At the end of this study, various numerical methods, which applied in the different electrospinning simulations, were revised.

KEYWORDS

- **electrospinning**
- **electrospinning modeling**
- **electrospinning simulation**
- **nanomaterial**
- **viscoelastic polymers**

REFERENCES

1. Poole, C.P. and F.J. Owens, *Introduction to Nanotechnology*. 2003, New Jersey, Hoboken: Wiley. 400.
2. Nalwa, H.S., *Nanostructured Materials and Nanotechnology: Concise Edition*. 2001: Gulf Professional Publishing.
3. Gleiter, H., *Nanostructured Materials: State of the Art and Perspectives*. Nanostructured Materials, 1995. 6(1): p. 3–14.
4. Wong, Y., et al., *Selected Applications of Nanotechnology in Textiles*. AUTEX Research Journal, 2006. 6(1): p. 1–8.
5. Yu, B. and M. Meyyappan, *Nanotechnology: Role in Emerging Nanoelectronics*. Solid-state electronics, 2006. 50(4): p. 536–544.
6. Farokhzad, O.C. and R. Langer, *Impact of Nanotechnology on Drug Delivery*. ACS Nano, 2009. 3(1): p. 16–20.
7. Serrano, E., G. Rus, and J. Garcia-Martinez, *Nanotechnology for Sustainable Energy*. Renewable and Sustainable Energy Reviews, 2009. 13(9): p. 2373–2384.
8. Dreher, K.L., *Health and Environmental Impact of Nanotechnology: Toxicological Assessment of Manufactured Nanoparticles*. Toxicological Sciences, 2004. 77(1): p. 3–5.
9. Bhushan, B., *Introduction to Nanotechnology*, in *Springer Handbook of Nanotechnology*. 2010, Springer. p. 1–13.
10. Ratner, D. and M.A. Ratner, *Nanotechnology and Homeland Security: New Weapons for New Wars*. 2004: Prentice Hall Professional. 145.
11. Aricò, A.S., et al., *Nanostructured Materials for Advanced Energy Conversion and Storage Devices*. Nature Materials, 2005. 4(5): p. 366–377.
12. Wang, Z.L., *Nanomaterials for Nanoscience and Nanotechnology*. Characterization of Nanophase Materials, 2000: p. 1–12.
13. Gleiter, H., *Nanostructured Materials: Basic Concepts and Microstructure*. Acta Materialia, 2000. 48(1): p. 1–29.
14. Wang, X., et al., *A General Strategy for Nanocrystal Synthesis*. Nature, 2005. 437(7055): p. 121–124.
15. Kelsall, R.W., et al., *Nanoscale Science and Technology*. 2005, New York: Wiley Online Library.
16. Engel, E., et al., *Nanotechnology in Regenerative Medicine: The Materials Side*. Trends in Biotechnology, 2008. 26(1): p. 39–47.
17. Beachley, V. and X. Wen, *Polymer Nanofibrous Structures: Fabrication, Biofunctionalization, and Cell interactions*. Progress in Polymer Science, 2010. 35(7): p. 868–892.
18. Gogotsi, Y., *Nanomaterials Handbook*. 2006, New York: CRC press. 779.
19. Li, C. and T. Chou, *A Structural Mechanics Approach for the Analysis of Carbon Nanotubes*. International Journal of Solids and Structures, 2003. 40(10): p. 2487–2499.
20. Delerue, C. and M. Lannoo, *Nanostructures: Theory and Modeling*. 2004: Springer. 304.
21. Pokropivny, V. and V. Skorokhod, *Classification of Nanostructures by Dimensionality and Concept of Surface Forms Engineering in Nanomaterial Science*. Materials Science and Engineering: C, 2007. 27(5): p. 990–993.
22. Balbuena, P. and J.M. Seminario, *Nanomaterials: Design and Simulation: Design and Simulation*. Vol. 18. 2006: Elsevier.

23. Kawaguchi, T. and H. Matsukawa, *Numerical Study of Nanoscale Lubrication and Friction at Solid Interfaces.* Molecular Physics, 2002. 100(19): p. 3161–3166.
24. Ponomarev, S.Y., K.M. Thayer, and D.L. Beveridge, *Ion Motions in Molecular Dynamics Simulations on DNA.* Proceedings of the National Academy of Sciences of the United States of America, 2004. 101(41): p. 14771–14775.
25. Loss, D. and D.P. DiVincenzo, *Quantum computation with quantum dots.* Physical Review A, 1998. 57(1): p.120–125.
26. Theodosiou, T.C. and D.A. Saravanos, *Molecular Mechanics Based Finite Element for Carbon Nanotube Modeling.* Computer Modeling In Engineering And Sciences, 2007. 19(2): p. 19–24.
27. Pokropivny, V. and V. Skorokhod, *New Dimensionality Classifications of Nanostructures.* Physica E: Low-dimensional Systems and Nanostructures, 2008. 40(7): p. 2521–2525.
28. Lieber, C.M., *One-dimensional Nanostructures: Chemistry, Physics & Applications.* Solid State Communications, 1998. 107(11): p. 607–616.
29. Emary, C., *Theory of Nanostructures.* 2009.
30. Edelstein, A.S. and R.C. Cammaratra, *Nanomaterials: Synthesis, Properties and Applications.* 1998: CRC Press.
31. Grzelczak, M., et al., *Directed Self-assembly of Nanoparticles.* ACS Nano, 2010. 4(7): p. 3591–3605.
32. Hung, C., et al., *Strain Directed Assembly of Nanoparticle Arrays within a Semiconductor.* Journal of Nanoparticle Research, 1999. 1(3): p. 329–347.
33. Wang, L. and R. Hong, *Synthesis, Surface Modification and Characterization of Nanoparticles.* Polymer Composites. 2: p. 13–51.
34. Lai, W., et al., *Synthesis of Nanostructured Materials by Hot and Cold Plasma.*
35. Petermann, N., et al., *Plasma Synthesis of Nanostructures for Improved Thermoelectric Properties.* Journal of Physics D: Applied Physics, 2011. 44(17): p. 174034.
36. Ye, Y., et al., *RF plasma method.* 2001, Google Patents.
37. Hyeon, T., *Chemical Synthesis of Magnetic Nanoparticles.* Chemical Communications, 2003(8): p. 927–934.
38. Galvez, A., et al., *Carbon Nanoparticles from Laser Pyrolysis.* Carbon, 2002. 40(15): p. 2775–2789.
39. Porterat, D., *Synthesis of Nanoparticles by Laser Pyrolysis.* 2012, Google Patents.
40. Tiwari, J.N., R.N. Tiwari, and K.S. Kim, *Zero-dimensional, One-dimensional, Two-dimensional and Three-dimensional Nanostructured Materials for Advanced Electrochemical Energy Devices.* Progress in Materials Science, 2012. 57(4): p. 724–803.
41. Murray, P.T., et al. *Nanomaterials Produced by Laser Ablation Techniques Part I: Synthesis and Passivation of Nanoparticles.* in *Nondestructive Evaluation for Health Monitoring and Diagnostics.* 2006: International Society for Optics and Photonics.
42. Dolgaev, S.I., et al., *Nanoparticles Produced by Laser Ablation of Solids in Liquid Environment.* Applied Surface Science, 2002. 186(1): p. 546–551.
43. Becker, M.F., et al., *Metal Nanoparticles Generated by Laser Ablation.* Nanostructured Materials, 1998. 10(5): p. 853–863.
44. Bonneau, F., et al., *Numerical Simulations for Description of UV Laser Interaction with Gold Nanoparticles Embedded in Silica.* Applied Physics B, 2004. 78(3–4): p. 447–452.

45. Chen, Y.H. and C.S. Yeh, *Laser Ablation Method: Use of Surfactants to Form the Dispersed Ag Nanoparticles.* Colloids and Surfaces A: Physicochemical and Engineering Aspects, 2002. 197(1): p. 133–139.
46. Andrady, A.L., *Science and Technology of Polymer Nanofibers*. 2008, Hoboken: John Wiley & Sons, Inc. 404.
47. Wang, H.S., G.D. Fu, and X.S. Li, *Functional Polymeric Nanofibers From Electrospinning.* Recent Patents on Nanotechnology, 2009. 3(1): p. 21–31.
48. Ramakrishna, S., *An Introduction to Electrospinning and Nanofibers*. 2005: World Scientific Publishing Company. 396.
49. Reneker, D.H. and I. Chun, *Nanometer Diameter Fibres of Polymer, Produced by Electrospinning.* Nanotechnology, 1996. 7(3): p. 216.
50. Doshi, J. and D.H. Reneker, *Electrospinning Process and Applications of Electrospun Fibers.* Journal of Electrostatics, 1995. 35(2): p. 151–160.
51. Burger, C., B. Hsiao, and B. Chu, *Nanofibrous Materials and Their Applications.* Annual Reviews Material Researchs, 2006. 36: p. 333–368.
52. Fang, J., et al., *Applications of Electrospun Nanofibers.* Chinese Science Bulletin, 2008. 53(15): p. 2265–2286.
53. Ondarcuhu, T. and C. Joachim, *Drawing a Single Nanofiber Over Hundreds of Microns.* EPL (Europhysics Letters), 1998. 42(2): p. 215.
54. Nain, A.S., et al., *Drawing Suspended Polymer Micro/Nanofibers Using Glass Micropipettes.* Applied Physics Letters, 2006. 89(18): p. 183105–183105-3.
55. Bajakova, J., et al., *"Drawing"-The Production of Individual Nanofibers by Experimental Method*, in *Nanoconference*. 2011: Brno, Czech Republic, EU.
56. Feng, L., et al., *Super Hydrophobic Surface of Aligned Polyacrylonitrile Nanofibers.* Angewandte Chemie, 2002. 114(7): p. 1269–1271.
57. Delvaux, M., et al., *Chemical and Electrochemical Synthesis of Polyaniline Micro-and Nano-tubules.* Synthetic Metals, 2000. 113(3): p. 275–280.
58. Barnes, C.P., et al., *Nanofiber Technology: Designing the Next Generation of Tissue Engineering Scaffolds.* Advanced Drug Delivery Reviews, 2007. 59(14): p. 1413–1433.
59. Palmer, L.C. and S.I. Stupp, *Molecular Self-assembly into One-dimensional Nanostructures.* Accounts of Chemical Research, 2008. 41(12): p. 1674–1684.
60. Hohman, M.M., et al., *Electrospinning and Electrically Forced jets. I. Stability Theory.* Physics of Fluids, 2001. 13: p. 2201–2220.
61. Hohman, M.M., et al., *Electrospinning and Electrically Forced Jets. II. Applications.* Physics of Fluids, 2001. 13: p. 2221.
62. Shin, Y.M., et al., *Experimental Characterization of Electrospinning: The Electrically Forced Jet and Instabilities.* Polymer, 2001. 42(25): p. 9955–9967.
63. Fridrikh, S.V., et al., *Controlling the fiber diameter during electrospinning.* Physical review letters, 2003. 90(14): p. 144502–144502.
64. Yarin, A.L., S. Koombhongse, and D.H. Reneker, *Taylor Cone and Jetting from Liquid droplets in Electrospinning of Nanofibers.* Journal of Applied Physics, 2001. 90(9): p. 4836–4846.
65. Zeleny, J., *The electrical discharge from liquid points, and a hydrostatic method of measuring the electric intensity at their surfaces.* Physical Review, 1914. 3(2): p. 69–91.

66. Reneker, D.H., et al., *Bending Instability of Electrically Charged Liquid Jets of Polymer Solutions in Electrospinning.* Journal of Applied Physics, 2000. 87: p. 4531–4547.
67. Frenot, A. and I.S. Chronakis, *Polymer Nanofibers Assembled by Electrospinning.* Current Opinion in Colloid & Interface Science, 2003. 8(1): p. 64–75.
68. Gilbert, W., *De Magnete* Transl. PF Mottelay, Dover, UK. 1958, New York: Dover Publications, Inc. 366.
69. Tucker, N., et al., *The History of the Science and Technology of Electrospinning from 1600 to 1995.* Journal of Engineered Fibers and Fabrics, 2012. 7: p. 63–73.
70. Hassounah, I., *Melt electrospinning of thermoplastic polymers*. 2012: Aachen: Hochschulbibliothek Rheinisch-Westfälische Technischen Hochschule Aachen. 650.
71. Taylor, G.I., *The Scientific Papers of Sir Geoffrey Ingram Taylor.* Mechanics of Fluids, 1971. 4.
72. Yeo, L.Y. and J.R. Friend, *Electrospinning Carbon Nanotube Polymer Composite Nanofibers.* Journal of Experimental Nanoscience, 2006. 1(2): p. 177–209.
73. Bhardwaj, N. and S.C. Kundu, *Electrospinning: a fascinating fiber fabrication technique.* Biotechnology Advances, 2010. 28(3): p. 325–347.
74. Huang, Z.M., et al., *A review on polymer nanofibers by electrospinning and their applications in nanocomposites.* Composites Science and Technology, 2003. 63(15): p. 2223–2253.
75. Haghi, A.K., *Electrospinning of nanofibers in textiles*. 2011, North Carolina: Apple Academic Press Inc. 132.
76. Bhattacharjee, P., V. Clayton, and A.G. Rutledge, *Electrospinning and Polymer Nanofibers: Process Fundamentals*, in *Comprehensive Biomaterials*. 2011, Elsevier. p. 497–512.
77. Garg, K. and G.L. Bowlin, *Electrospinning jets and nanofibrous structures*. Biomicrofluidics, 2011. 5: p. 13403–13421.
78. Angammana, C.J. and S.H. Jayaram, *A Theoretical Understanding of the Physical Mechanisms of Electrospinning*, in *Proc. ESA Annual Meeting on Electrostatics*. 2011: Case Western Reserve University, Cleveland OH. p. 1–9.
79. Reneker, D.H. and A.L. Yarin, *Electrospinning Jets and Polymer Nanofibers.* Polymer, 2008. 49(10): p. 2387–2425.
80. Deitzel, J., et al., *The Effect of Processing Variables on the Morphology of Electrospun Nanofibers and Textiles.* Polymer, 2001. 42(1): p. 261–272.
81. Rutledge, G.C. and S.V. Fridrikh, *Formation of Fibers by Electrospinning.* Advanced Drug Delivery Reviews, 2007. 59(14): p. 1384–1391.
82. De Vrieze, S., et al., *The Effect of Temperature and Humidity on Electrospinning.* Journal of Materials Science, 2009. 44(5): p. 1357–1362.
83. Kumar, P., *Effect of Collector on Electrospinning to Fabricate Aligned Nanofiber*, in *Department of Biotechnology and Medical Engineering*. 2012, National Institute of Technology Rourkela: Rourkela.
84. Reneker, D.H., et al., *Electrospinning of Nanofibers from Polymer Solutions and Melts.* Advances in Applied Mechanics, 2007. 41: p. 343–346.
85. Haghi, A.K. and G. Zaikov, *Advances in Nanofiber Research.* 2012: Smithers Rapra Technology. 194.
86. Maghsoodloo, S., et al., *A Detailed Review on Mathematical Modeling of Electrospun Nanofibers.* Polymers Research Journal 2012. 6: p. 361–379.

87. Fritzson, P., *Principles of object-oriented modeling and simulation with Modelica 2.1*. 2010: Wiley-IEEE Press.
88. Collins, A.J., et al., *The Value of Modeling and Simulation Standards*. 2011, Virginia Modeling, Analysis and Simulation Center, Old Dominion University: Virginia. p. 1–8.
89. Robinson, S., *Simulation: the practice of model development and use*. 2004: Wiley. 722.
90. Carson, I.I. and S. John, *Introduction to modeling and simulation*, in *Proceedings of the 36th conference on Winter simulation*. 2004, Winter Simulation Conference: Washington, DC. p. 9–16.
91. Banks, J., *Handbook of simulation*. 1998: Wiley Online Library. 342.
92. Pritsker, A.B. and B. Alan, *Principles of Simulation Modeling*. 1998, New York: Wiley. 426.
93. Carroll, C.P., *The development of a comprehensive simulation model for electrospinning*. Vol. 70. 2009, Cornell University 300.
94. Carroll, C.P., et al., *Nanofibers From Electrically Driven Viscoelastic Jets: Modeling and Experiments*. Korea-Australia Rheology Journal, 2008. 20(3): p. 153–164.
95. Yu, J.H., S.V. Fridrikh, and G.C. Rutledge, *The Role of Elasticity in the Formation of Electrospun Fibers*. Polymer, 2006. 47(13): p. 4789–4797.
96. Han, T., A.L. Yarin, and D.H. Reneker, *Viscoelastic Electrospun Jets: Initial Stresses and Elongational Rheometry*. Polymer, 2008. 49(6): p. 1651–1658.
97. Bhattacharjee, P.K., et al., *Extensional Stress Growth and Stress Relaxation in Entangled Polymer Solutions*. Journal of Rheology, 2003. 47: p. 269–290.
98. Paruchuri, S. and M.P. Brenner, *Splitting of A Liquid Jet*. Physical Review Letters, 2007. 98(13): p. 134502–134504.
99. Ganan-Calvo, A.M., *On the Theory of Electrohydrodynamically Driven Capillary Jets*. Journal of Fluid Mechanics, 1997. 335: p. 165–188.
100. Liu, L. and Y.A. Dzenis, *Simulation of Electrospun Nanofiber Deposition on Stationary and Moving Substrates*. Micro & Nano Letters, 2011. 6(6): p. 408–411.
101. Spivak, A.F. and Y.A. Dzenis, *Asymptotic decay of radius of a weakly conductive viscous jet in an external electric field*. Applied Physics Letters, 1998. 73(21): p. 3067–3069.
102. Jaworek, A. and A. Krupa, *Classification of the Modes of EHD Spraying*. Journal of Aerosol Science, 1999. 30(7): p. 873–893.
103. Senador, A.E., M.T. Shaw, and P.T. Mather. *Electrospinning of Polymeric Nanofibers: Analysis of jet formation*. in *MRS Proceedings*. 2000: Cambridge University Press.
104. Feng, J.J., *The stretching of an electrified nonNewtonian jet: A model for electrospinning*. Physics of Fluids, 2002. 14(11): p. 3912–3927.
105. Feng, J.J., *Stretching of a straight electrically charged viscoelastic jet*. Journal of Non-Newtonian Fluid Mechanics, 2003. 116(1): p. 55–70.
106. Spivak, A.F., Y.A. Dzenis, and D.H. Reneker, *A Model of Steady State Jet in the Electrospinning Process*. Mechanics Research Communications, 2000. 27(1): p. 37–42.
107. Yarin, A.L., S. Koombhongse, and D.H. Reneker, *Bending Instability in Electrospinning of Nanofibers*. Journal of Applied Physics, 2001. 89: p. 3018.

108. Gradoń, L., *Principles of Momentum, Mass and Energy Balances.* Chemical Engineering and Chemical Process Technology. 1: p. 1–6.
109. Bird, R.B., W.E. Stewart, and E.N. Lightfoot, *Transport Phenomena.* Vol. 2. 1960, New York: Wiley & Sons, Incorporated, John 808.
110. Peters, G.W.M., M.A. Hulsen, and R.H.M. Solberg, *A Model for Electrospinning Viscoelastic Fluids*, in *Department of Mechanical Engineering.* 2007, Eindhoven University of Technology: Eindhoven p. 26.
111. Whitaker, R.D., *An historical note on the conservation of mass.* Journal of Chemical Education, 1975. 52(10): p. 658.
112. He, J.H., et al., *Mathematical models for continuous electrospun nanofibers and electrospun nanoporous microspheres.* Polymer International, 2007. 56(11): p. 1323–1329.
113. Xu, L., F. Liu, and N. Faraz, *Theoretical model for the electrospinning nanoporous materials process.* Computers and Mathematics with Applications, 2012. 64(5): p. 1017–1021.
114. Heilbron, J.L., *Electricity in the 17th and eighteenth century: A Study of Early Modern Physics.* 1979: University of California Press. 437.
115. Orito, S. and M. Yoshimura, *Can the universe be charged?* Physical Review Letters, 1985. 54(22): p. 2457–2460.
116. Karra, S., *Modeling electrospinning process and a numerical scheme using Lattice Boltzmann method to simulate viscoelastic fluid flows*, in *Indian Institute of Technology.* 2007, Texas A&M University: Chennai. p. 60.
117. Hou, S.H. and C.K. Chan, *Momentum Equation for Straight Electrically Charged Jet.* Applied Mathematics and Mechanics, 2011. 32(12): p. 1515–1524.
118. Maxwell, J.C., *Electrical Research of the Honorable Henry Cavendish, 426*, in *Cambridge University Press*, Cambridge, Editor. 1878, Cambridge University Press, Cambridge, UK: UK.
119. Vught, R.V., *Simulating the dynamical behavior of electrospinning processes*, in *Department of Mechanical Engineering.* 2010, Eindhoven University of Technology: Eindhoven. p. 68.
120. Jeans, J.H., *The Mathematical Theory of Electricity and Magnetism.* 1927, London: Cambridge University Press. 536.
121. Truesdell, C. and W. Noll, *The nonlinear field theories of mechanics.* 2004: Springer. 579.
122. Roylance, D., *Constitutive equations*, in *Lecture Notes. Department of Materials Science and Engineering.* 2000, Massachusetts Institute of Technology: Cambridge. p. 10.
123. He, J.H., Y. Wu, and N. Pang, *A mathematical model for preparation by AC-electrospinning process.* International Journal of Nonlinear Sciences and Numerical Simulation, 2005. 6(3): p. 243–248.
124. Little, R.W., *Elasticity.* 1999: Courier Dover Publications. 431.
125. Clauset, A., C.R. Shalizi, and M.E.J. Newman, *Power-law Distributions in Empirical Data.* SIAM Review, 2009. 51(4): p. 661–703.
126. Wan, Y., Q. Guo, and N. Pan, *Thermo-electro-hydrodynamic model for electrospinning process.* International Journal of Nonlinear Sciences and Numerical Simulation, 2004. 5(1): p. 5–8.

127. Giesekus, H., *Die elastizität von flüssigkeiten.* Rheologica Acta, 1966. 5(1): p. 29–35.
128. Giesekus, H., *The physical meaning of Weissenberg's hypothesis with regard to the second normal-stress difference*, in *The Karl Weissenberg 80th Birthday Celebration Essays*, J. Harris and K. Weissenberg, Editors. 1973, East African Literature Bureau p. 103–112.
129. Wiest, J.M., *A differential constitutive equation for polymer melts.* Rheologica Acta, 1989. 28(1): p. 4–12.
130. Bird, R.B. and J.M. Wiest, *Constitutive Equations for Polymeric Liquids.* Annual Review of Fluid Mechanics, 1995. 27(1): p. 169–193.
131. Giesekus, H., *A simple constitutive equation for polymer fluids based on the concept of deformation-dependent tensorial mobility.* Journal of Non-Newtonian Fluid Mechanics, 1982. 11(1): p. 69–109.
132. Oliveira, P.J., *On the Numerical Implementation of Nonlinear Viscoelastic Models in a Finite-Volume Method.* Numerical Heat Transfer: Part B: Fundamentals, 2001. 40(4): p. 283–301.
133. Simhambhatla, M. and A.I. Leonov, *On the Rheological Modeling of Viscoelastic Polymer Liquids with Stable Constitutive Equations.* Rheologica Acta, 1995. 34(3): p. 259–273.
134. Giesekus, H., *A unified approach to a variety of constitutive models for polymer fluids based on the concept of configuration-dependent molecular mobility.* Rheologica Acta, 1982. 21(4–5): p. 366–375.
135. Eringen, A.C. and G.A. Maugin, *Electrohydrodynamics*, in *Electrodynamics of Continua II.* 1990, Springer. p. 551–573.
136. Hutter, K., *Electrodynamics of Continua (A. Cemal Eringen and Gerard A. Maugin).* SIAM Review, 1991. 33(2): p. 315–320.
137. Kröger, M., *Simple Models for Complex Nonequilibrium Fluids.* Physics Reports, 2004. 390(6): p. 453–551.
138. Denn, M.M., *Issues in Viscoelastic Fluid Mechanics.* Annual Review of Fluid Mechanics, 1990. 22(1): p. 13–32.
139. Rossky, P.J., J.D. Doll, and H.L. Friedman, *Brownian Dynamics as Smart Monte Carlo Simulation.* The Journal of Chemical Physics, 1978. 69: p. 4628–4633.
140. Chen, J.C. and A.S. Kim, *Brownian Dynamics, Molecular Dynamics, and Monte Carlo Modeling of Colloidal Systems.* Advances in Colloid and Interface Science, 2004. 112(1): p. 159–173.
141. Pasini, P. and C. Zannoni, *Computer Simulations of Liquid Crystals and Polymers.* Vol. 177. 2005, Erice: Springer. 380.
142. Zhang, H. and P. Zhang, *Local Existence for the FENE-dumbbell Model of Polymeric Fluids.* Archive for Rational Mechanics and Analysis, 2006. 181(2): p. 373–400.
143. Isihara, A., *Theory of High Polymer Solutions (The Dumbbell Model).* The Journal of Chemical Physics, 1951. 19: p. 397–343.
144. Masmoudi, N., *Well-posedness for the FENE Dumbbell Model of Polymeric Flows.* Communications on Pure and Applied Mathematics, 2008. 61(12): p. 1685–1714.
145. Stockmayer, W.H., et al., *Dynamic Properties of Solutions. Models for Chain Molecule Dynamics in Dilute Solution.* Discussions of the Faraday Society, 1970. 49: p. 182–192.

146. Graham, R.S., et al., *Microscopic Theory of Linear, Entangled Polymer Chains under Rapid Deformation Including Chain Stretch and Convective Constraint Release.* Journal of Rheology, 2003. 47: p. 1171–1200.
147. Gupta, R.K., E. Kennel, and K.S. Kim, *Polymer nanocomposites handbook.* 2010: CRC Press.
148. Marrucci, G., *The free energy constitutive equation for polymer solutions from the dumbbell model.* Journal of Rheology, 1972. 16: p. 321–331.
149. Reneker, D.H., et al., *Electrospinning of Nanofibers from Polymer Solutions and Melts.* Advances in Applied Mechanics, 2007. 41: p. 43–195.
150. Kowalewski, T.A., S. Barral, and T. Kowalczyk, *Modeling Electrospinning of Nanofibers*, in *IUTAM Symposium on Modeling Nanomaterials and Nanosystems*. 2009, Springer: Aalborg, Denmark. p. 279–292.
151. Macosko, C.W., *Rheology: Principles, Measurements, and Applications.* Poughkeepsie,. 1994, New york: Wiley-VCH. 578.
152. Kowalewski, T.A., S. Blonski, and S. Barral, *Experiments and Modeling of Electrospinning Process.* Technical Sciences, 2005. 53(4): p. 385–394.
153. Song, Y.S. and J.R. Youn, *Modeling of Rheological Behavior of Nanocomposites by Brownian Dynamics Simulation.* Korea-Australia Rheology Journal, 2004. 16(4): p. 201–212.
154. Ma, W.K.A., et al., *Rheological Modeling of Carbon Nanotube Aggregate Suspensions.* Journal of Rheology, 2008. 52: p. 1311–1330.
155. Buysse, W.M., *A 2D Model for the Electrospinning Process*, in *Department of Mechanical Engineering*. 2008, Eindhoven University of Technology: Eindhoven. p. 71.
156. Silling, S.A. and F. Bobaru, *Peridynamic Modeling of Membranes and Fibers.* International Journal of Non-Linear Mechanics, 2005. 40(2): p. 395–409.
157. Teo, W.E. and S. Ramakrishna, *Electrospun Nanofibers as a Platform for Multifunctional, Hierarchically Organized Nanocomposite.* Composites Science and Technology, 2009. 69(11): p. 1804–1817.
158. Wu, X. and Y.A. Dzenis, *Elasticity of Planar Fiber Networks.* Journal of Applied Physics, 2005. 98(9): p. 93501.
159. Tatlier, M. and L. Berhan, *Modeling the Negative Poisson's Ratio of Compressed Fused Fiber Networks.* Physica Status Solidi (b), 2009. 246(9): p. 2018–2024.
160. Kuipers, B., *Qualitative reasoning: modeling and simulation with incomplete knowledge*. 1994: the MIT press. 554.
161. West, B.J., *Comments on the renormalization group, scaling and measures of complexity.* Chaos, Solitons and Fractals, 2004. 20(1): p. 33–44.
162. De Gennes, P.G. and T.A. Witten, *Scaling Concepts in Polymer Physics*. Vol. Cornell University Press. 1980. 324.
163. He, J.H. and H.M. Liu, *Variational approach to nonlinear problems and a review on mathematical model of electrospinning.* Nonlinear Analysis, 2005. 63: p. e919-e929.
164. He, J.H., Y.Q. Wan, and J.Y. Yu, *Allometric scaling and instability in electrospinning.* International Journal of Nonlinear Sciences and Numerical Simulation 2004. 5(3): p. 243–252.
165. He, J.H., Y.Q. Wan, and J.Y. Yu, *Allometric Scaling and Instability in Electrospinning.* International Journal of Nonlinear Sciences and Numerical Simulation, 2004. 5: p. 243–252.

166. He, J.H. and Y.Q. Wan, *Allometric scaling for voltage and current in electrospinning.* Polymer, 2004. 45: p. 6731–6734.
167. He, J.H., Y.Q. Wan, and J.Y. Yu, *Scaling law in electrospinning: relationship between electric current and solution flow rate.* Polymer, 2005. 46: p. 2799–2801.
168. He, J.H., Y.Q. Wanc, and J.Y. Yuc, *Application of vibration technology to polymer electrospinning.* International Journal of Nonlinear Sciences and Numerical Simulation, 2004. 5(3): p. 253–262.
169. Kessick, R., J. Fenn, and G. Tepper, *The use of AC potentials in electrospraying and electrospinning processes.* Polymer, 2004. 45(9): p. 2981–2984.
170. Boucher, D.F. and G.E. Alves, *Dimensionless numbers, part 1 and 2.* 1959.
171. Ipsen, D.C., *Units Dimensions And Dimensionless Numbers.* 1960, New York: McGraw Hill Book Company Inc. 466.
172. Langhaar, H.L., *Dimensional analysis and theory of models.* Vol. 2. 1951, New York: Wiley. 166
173. McKinley, G.H., *Dimensionless groups for understanding free surface flows of complex fluids.* Bulletin of the Society of Rheology, 2005. 2005: p. 6–9.
174. Carroll, C.P., et al., *Nanofibers from Electrically Driven Viscoelastic Jets: Modeling and Experiments.* Korea-Australia Rheology Journal, 2008. 20(3): p. 153–164.
175. Saville, D.A., *Electrohydrodynamics: the Taylor-Melcher leaky dielectric model.* Annual Review of Fluid Mechanics, 1997. 29(1): p. 27–64.
176. Ramos, J.I., *Force Fields on Inviscid, Slender, Annular Liquid.* International Journal for Numerical Methods in Fluids, 1996. 23: p. 221–239.
177. Ha, J.W. and S.M. Yang, *Deformation and breakup of Newtonian and nonNewtonian conducting drops in an electric field.* Journal of Fluid Mechanics, 2000. 405: p. 131–156.
178. Reneker, D.H., et al., *Bending Instability of Electrically Charged Liquid Jets of Polymer Solutions in Electrospinning.* Journal of Applied physics, 2000. 87: p. 4531.
179. Peters, G., M. Hulsen, and R. Solberg, *A Model for Electrospinning Viscoelastic Fluids.*
180. Wan, Y., et al., *Modeling and Simulation of the Electrospinning Jet with Archimedean Spiral.* Advanced Science Letters, 2012. 10(1): p. 590–592.
181. Dasri, T., *Mathematical Models of Bead-Spring Jets during Electrospinning for Fabrication of Nanofibers.* Walailak Journal of Science and Technology, 2012. 9.
182. Solberg, R.H.M., *Position-controlled deposition for electrospinning.* 2007, Eindhoven University of Technology: Eindhoven. p. 75.
183. Holzmeister, A., A.L. Yarin, and J.H. Wendorff, *Barb Formation in Electrospinning: Experimental and Theoretical Investigations.* Polymer, 2010. 51(12): p. 2769–2778.
184. Karra, S., *Modeling electrospinning process and a numerical scheme using Lattice Boltzmann method to simulate viscoelastic fluid flows*. 2012.
185. Arinstein, A., et al., *Effect of supramolecular structure on polymer nanofiber elasticity.* Nature Nanotechnology, 2007. 2(1): p. 59–62.
186. Lu, C., et al., *Computer Simulation of Electrospinning. Part I. Effect of Solvent in Electrospinning.* Polymer, 2006. 47(3): p. 915–921.
187. Greenfeld, I., et al., *Polymer dynamics in semidilute solution during electrospinning: A simple model and experimental observations.* Physical Review 2011. 84(4): p. 41806–41815.

188. Ly, H.V. and H.T. Tran, *Modeling and Control of Physical Processes Using Proper Orthogonal Decomposition.* Mathematical and Computer Modeling, 2001. 33(1): p. 223–236.
189. Peiró, J. and S. Sherwin, *Finite Difference, Finite Element and Finite Volume Methods for Partial Differential Equations*, in *Handbook of Materials Modeling.* 2005, Springer: London. p. 2415–2446.
190. Kitano, H., *Computational Systems Biology.* Nature, 2002. 420(6912): p. 206–210.
191. Gerald, C.F. and P.O. Wheatley, *Applied Numerical Analysis*, ed. 7th. 2007: Addison-Wesley. 624
192. Burden, R.L. and J.D. Faires, *Numerical Analysis.* Vol. 8. 2005: Thomson Brooks/ Cole. 850.
193. Lawrence, C.E., *Partial Differential Equations.* 2010: American Mathematical Society. 749.
194. Quarteroni, A., A.M. Quarteroni, and A. Valli, *Numerical Approximation of Partial Differential Equations.* Vol. 23. 2008: Springer. 544.
195. Butcher, J.C., *A History of Runge-Kutta Methods.* Applied Numerical Mathematics, 1996. 20(3): p. 247–260.
196. Cartwright, J.H.E. and O. Piro, *The Dynamics of Runge–Kutta Methods.* International Journal of Bifurcation and Chaos, 1992. 2(03): p. 427–449.
197. Zingg, D.W. and T.T. Chisholm, *Runge–Kutta Methods for Linear Ordinary Differential Equations.* Applied Numerical Mathematics, 1999. 31(2): p. 227–238.
198. Butcher, J.C., *The Numerical Analysis of Ordinary Differential Equations: Runge-Kutta and General Linear Methods.* 1987: Wiley-Interscience. 512.
199. Reznik, S.N., et al., *Evolution of a Compound Droplet Attached to a Core shell Nozzle Under the Action of a Strong Electric Field.* Physics of Fluids, 2006. 18(6): p. 062101–062101–13.
200. Reznik, S.N., et al., *Transient and Steady Shapes of Droplets Attached to a Surface in a Strong Electric Field.* Journal of Fluid Mechanics, 2004. 516: p. 349–377.
201. Donea, J. and A. Huerta, *Finite Element Methods for Flow Problems.* 2003: Wiley. com. 362.
202. Zienkiewicz, O.C. and R.L. Taylor, *The Finite Element Method: Solid Mechanics.* Vol. 2. 2000: Butterworth-heinemann. 459.
203. Brenner, S.C. and L.R. Scott, *The Mathematical Theory of Finite Element Methods.* Vol. 15. 2008: Springer. 397.
204. Bathe, K.J., *Finite Element Procedures.* Vol. 2. 1996: Prentice hall Englewood Cliffs. 1037.
205. Reddy, J.N., *An Introduction to the Finite Element Method.* Vol. 2. 2006: McGraw-Hill New York. 912.
206. Ferziger, J.H. and M. Perić, *Computational Methods for Fluid Dynamics.* Vol. 3. 1996: Springer Berlin. 423.
207. P.T. Baaijens, F., *Mixed Finite Element Methods for Viscoelastic Flow Analysis: A Review.* Journal of Non-Newtonian Fluid Mechanics, 1998. 79(2): p. 361–385.
208. Angammana, C.J. and S.H. Jayaram. *A Theoretical Understanding of the Physical Mechanisms of Electrospinning.* in *Proceedings of the ESA Annual Meeting on Electrostatics.* 2011.

209. Costabel, M., *Principles of Boundary Element Methods.* Computer Physics Reports, 1987. 6(1): p. 243–274.
210. Kurz, S., J. Fetzer, and G. Lehner, *An Improved Algorithm for the BEM-FEM-coupling Method Using Domain Decomposition.* IEEE Transactions on Magnetics, 1995. 31(3): p. 1737–1740.
211. Mushtaq, M., N.A. Shah, and G. Muhammad, *Advantages and Disadvantages of Boundary Element Methods For Compressible Fluid Flow Problems.* Journal of American Science, 2010. 6(1): p. 162–165.
212. Gaul, L., M. Kögl, and M. Wagner, *Boundary Element Methods for Engineers and Scientists.* 2003: Springer. 488.
213. Kowalewski, T.A., S. Barral, and T. Kowalczyk. *Modeling Electrospinning of Nanofibers.* in *IUTAM Symposium on Modeling Nanomaterials and Nanosystems.* 2009: Springer.
214. Toro, E.F., *Riemann Solvers and Numerical Methods for Fluid Dynamics: A Practical Introduction.* 2009: Springer. 724.
215. Tonti, E., *A Direct Discrete Formulation of Field Laws: The Cell Method.* CMES-Computer Modeling in Engineering and Sciences, 2001. 2(2): p. 237–258.
216. Thomas, P.D. and C.K. Lombard, *Geometric Conservation Law and Its Application to Flow Computations on Moving Grids.* American Institute of Aeronautics and Astronautics Journal, 1979. 17(10): p. 1030–1037.
217. Lyrintzis, A.S., *Surface Integral Methods in Computational Aeroacoustics—From the (CFD) Near-field to the (Acoustic) Far-field.* International Journal of Aeroacoustics, 2003. 2(2): p. 95–128.
218. Škerget, L., M. Hriberšek, and G. Kuhn, *Computational Fluid Dynamics by Boundary–domain Integral Method.* International Journal for Numerical Methods in Engineering, 1999. 46(8): p. 1291–1311.
219. Rüberg, T. and F. Cirak, *An Immersed Finite Element Method with Integral Equation Correction.* International Journal for Numerical Methods in Engineering, 2011. 86(1): p. 93–114.
220. Feng, J.J., *The Stretching of an Electrified Non-Newtonian Jet: A Model for Electrospinning.* Physics of Fluids, 2002. 14: p. 3912–3926.
221. Varga, R.S., *Matrix Iterative Analysis.* Vol. 27. 2009: Springer. 358.
222. Stoer, J. and R. Bulirsch, *Introduction to Numerical Analysis.* Vol. 12. 2002: Springer. 744.
223. Bazaraa, M.S., H.D. Sherali, and C.M. Shetty, *Nonlinear Programming: Theory and Algorithms.* 2006: John Wiley & Sons. 872.
224. Fox, L., *Some Improvements in the Use of Relaxation Methods for the Solution of Ordinary and Partial Differential Equations.* Proceedings of the Royal Society of London. Series A. Mathematical and Physical Sciences, 1947. 190(1020): p. 31–59.
225. Zauderer, E., *Partial Differential Equations of Applied Mathematics.* Vol. 71. 2011: Wiley. com. 968.
226. Fisher, M.L., *The Lagrangian Relaxation Method for Solving Integer Programming Problems.* Management Science, 2004. 50(12 supplement): p. 1861–1871.
227. Steger, T.M., *Multi-Dimensional Transitional Dynamics: A Simple Numerical Procedure.* Macroeconomic Dynamics, 2005. 12(03): p. 301–319.

228. Roozemond, P.C., *A Model for Electrospinning Viscoelastic Fluids*, in *Department of Mechanical Engineering*. 2007, Eindhoven University of Technology. p. 25.
229. Succi, S., *The Lattice Boltzmann Equation: For Fluid Dynamics and Beyond*. 2001: Oxford university press. 288.
230. Chen, S. and G.D. Doolen, *Lattice Boltzmann Method for Fluid Flows*. Annual Review of Fluid Mechanics, 1998. 30(1): p. 329–364.
231. Aidun, C.K. and J.R. Clausen, *Lattice-Boltzmann Method for Complex Flows*. Annual Review of Fluid Mechanics, 2010. 42: p. 439–472.
232. Guo, Z. and T.S. Zhao, *Explicit Finite-difference Lattice Boltzmann Method for Curvilinear Coordinates*. Physical review E, 2003. 67(6): p. 066709–1-066709–12.
233. Albuquerque, P., et al., *A Hybrid Lattice Boltzmann Finite Difference Scheme for the Diffusion Equation*. International Journal for Multiscale Computational Engineering, 2006. 4(2): p. 209–219.
234. Tsutahara, M., *The Finite-difference Lattice Boltzmann Method and Its Application in Computational Aero-acoustics*. Fluid Dynamics Research, 2012. 44(4): p. 045507–1-045507–18.
235. Junk, M., *A Finite Difference Interpretation of the Lattice Boltzmann Method*. Numerical Methods for Partial Differential Equations, 2001. 17(4): p. 383–402.
236. So, R.M.C., S.C. Fu, and R.C.K. Leung, *Finite Difference Lattice Boltzmann Method for Compressible Thermal Fluids*. American Institute of Aeronautics and Astronautics Journal, 2010. 48(6): p. 1059–1071
237. Karra, S., *Modeling Electrospinning Process and a Numerical Scheme Using Lattice Boltzmann Method to Simulate Viscoelastic Fluid Flows*, in *Mechanical Engineering*. 2007, Indian Institute of Technology Madras. p. 60.
238. De Pascalis, R., *The Semi-Inverse Method in Solid Mechanics: Theoretical Underpinnings and Novel Applications*, in *Mathematics*. 2010, Universite Pierre et Marie Curie and Universita del Salento. p. 140.
239. Nemenyi, P.F., *Recent Developments in Inverse and Semi-inverse Methods in the Mechanics of Continua*. Advances in Applied Mechanics, 1951. 2(11): p. 123–151.
240. Chen, J.T., Y.T. Lee, and S.C. Shieh, *Revisit of Two Classical Elasticity Problems by Using the Trefftz Method*. Engineering Analysis with Boundary Elements, 2009. 33(6): p. 890–895.
241. Zhou, X.W. *A Note on the Semi-Inverse Method and a Variational Principle for the Generalized KdV-mKdV Equation*. in *Abstract and Applied Analysis*. 2013: Hindawi Publishing Corporation.
242. A. Narayan, S.P. and K.R. Rajagopal, *Unsteady Flows of a Class of Novel Generalizations of the Navier–Stokes Fluid*. Applied Mathematics and Computation, 2013. 219(19): p. 9935–9946.
243. He, J.H., *Variational Principles for Some Nonlinear Partial Differential Equations with Variable Coefficients*. Chaos, Solitons & Fractals, 2004. 19(4): p. 847–851.
244. Tarantola, A., *Inverse Problem Theory: Methods for Data Fitting and Model Parameter Estimation*. 2002: Elsevier Science. 613.
245. He, J.H., H.M. Liu, and N. Pan, *Variational Model for Ionomeric Polymer–metal Composite*. Polymer, 2003. 44(26): p. 8195–8199.
246. He, J.H., *Coupled Variational Principles of Piezoelectricity*. International Journal of Engineering Science, 2001. 39(3): p. 323–341.

247. Starovoitov, É. and F. Nağıyev, *Foundations of the Theory of Elasticity, Plasticity, and Viscoelasticity*. 2012: CRC Press. 320.
248. Bertero, M., *Regularization Methods for Linear Inverse Problems*, in *Inverse Problems*. 1986, Springer. p. 52–112.
249. Wan, Y.Q., Q. Guo, and N. Pan, *Thermo-electro-hydrodynamic Model for Eectrospinning Process*. International Journal of Nonlinear Sciences and Numerical Simulation, 2004. 5(1): p. 5–8.

CHAPTER 2

AFFINITY SEPARATION OF ENZYMES USING IMMOBILIZED METAL IONS PGMA GRAFTED CELLOPHANE MEMBRANES: β-GALACTOSIDASE ENZYME MODEL

M. S. MOHY ELDIN, M. A. ABU-SAIED, E.A. SOLIMAN, and E.A. HASSAN

CONTENTS

2.1 Introduction.......112
2.2 Experimental.......113
2.3 Results and Discussions.......117
2.4 Conclusions.......134
Keywords.......134
References.......135

2.1 INTRODUCTION

Immobilized Cu^{+2} ions affinity poly (glycidyl methacrylate) (PGMA) grafted cellophane membranes have been prepared, characterized and examined in the separation process of β-galactosidase (β-Gal) enzyme from its mixture with bovine serum albumin (BSA). The article focus on the step of Cu^{+2} ions immobilization onto sulfonated PGMA grafted membranes. Maximum amount of immobilized Cu^{+2} ions was found to be 60.9 ppm per g of polymer. The verification of the immobilization of Cu^{+2} ions step has been performed through characterization of the obtained membranes using EDAX analysis. Finally, Cu^{+2} immobilized membranes have been evaluated in separation of β-Gal enzyme from its mixture with BSA in different pH medium. Maximum protein adsorption, for both proteins, has been obtained at pH range 4–4.5; as 90 and 45% for β-Gal and BSA, respectively, from (1:20) mixture solution.

Immobilized metal ion affinity chromatography (IMAC), also called metal chelate chromatography, has been used widely for purification of proteins since its introduction [1] in 1975. Interactions of specific amino acid side chains, particularly those of histidine, cysteine, and tryptophan, with transition metal ions that are bound to the chelating groups of a chromatographic resin result in retention of proteins on the chromatographic column [2–7]. IMAC has been used to examine the relationship between amino acid side chain surface topography of proteins and binding selectivity [2]. One of the more recent applications of IMAC is the purification of recombinant proteins containing histidine tags (5, 7, 8, 9, 10). Among these techniques, IMAC is primarily based on affinity adsorption and therefore possesses advantages as well as disadvantages associated with this type of separation technology [11–16]. The advantages of IMAC, which include ligand stability, high protein loading, mild elution conditions, simple regeneration and low cost, are very important when developing protein purification procedures [17–22]. As reported earlier studies, the IDA ligand contains a secondary amino group and two carboxylic groups, which is a good ligand for immobilization of metal ions [23, 24]. Selection of polymers used for preparation of immobilized metal affinity support is an important factor that affect to the chromatographic performance of the support material. Gaberc Porekar and Menart, report that, the natural polymers may not suitable for large-scale industrial applications because of they exhibit low mechanical strength, inadequate porosity and large pressure drop [24]. These negative operational properties directed the researches toward the development of synthetic polymeric materials. Acrylate based polymers have good mechanical strength and can be prepared in various forms such as films, beads, rods, cryogels and flakes [25–29]. Membrane separation has attracted increasing attention for its potential capability in the field of separation, and it has been shown to be an effective technique in the chromatographic area. Microporous affinity films can significantly reduce mass transfer limitations commonly encountered in column chromatography [30–34]. As a conclusion, higher throughput, much lower operating pressure, and faster adsorption times are achieved in membrane systems [34–36]. Membrane separation is also a promising technology from the energy saving point of view for the selective separation of heavy metal ions [37–38].

In this chapter, Metals immobilization, selected membranes with highest sulphonation degree were immobilized with Cu^{+2} ions. The optimum conditions for the immobilization process are determined. The second part, deals with the membrane characterization. The effect of the grafting process on the hydrophilicity of the cellophane membranes has been monitored through changes in water uptake %. Two steps have been investigated in this part namely: Proteins adsorption, and Proteins elution. The studied parameters (pH, time of reaction, temperature change in the protein concentration) and affecting both steps have been evaluated and the optimum conditions are determined. The details of the obtained results and the proposed discussion are mentioned in the following and the immobilized Cu^{+2} ions membranes have been used in the affinity separation of β-Gal actosidase form protein mixture with BSA protein.

2.2 EXPERIMENTAL

2.2.1 MATERIALS

- Cellophane sheets were kindly supplied by Misr Rayon Co. Kafr El-Dawar (Egypt) type, uncoated; dimensions, 80 × 117 cm; cellulose content, 80% (W%) regenerated cellulose; additives content, 20% (glycerol and Na_2SiO_3). The additive were removed by extraction with hot distilled water, then the films were cut with dimensions 5 × 5 cm.

- Glycidyl methacrylate (GA) (Purity 97%) was obtained from Siga – Aldrich Chemicals Ltd. (Switzerland).
- Potassium persulfate (KPS) (Purity 99%, M.wt.270.31) was obtained from Siga-Aldrich Chemicals Ltd. (Germany).
- Ethyl alcohol absolute (Purity 99.9%) was obtained from El-Nasr Pharmaceutical Co. for Chemicals, (Egypt).
- Sodium sulphite anhydrous (Purity 95%) was obtained from El-Nasr Pharmaceutical Co. for Chemicals, (Egypt).
- Sulfuric acid (Purity 95–97%) was obtained from Siga- Aldrich Chemicals Ltd. (Germany).
- Methyl alcohol (Pure reagent for analysis) was obtained from El-Nasr Pharmaceutical Co. for Chemicals, (Egypt).
- 2-Propanol (Purity 99.8%) was obtained from Siga -Aldrich Chemicals Ltd. (Germany).
- Copper sulfate (Purity 98%, M.wt.249.68) was obtained from El-Nasr Pharmaceutical Co. for Chemicals, (Egypt).
- Hydrochloric acid (Purity 30–34%) was obtained from El-Nasr Pharmaceutical Co. for Chemicals, (Egypt).
- Bovine Serum Albumin (BSA Fraction V, minimum 96% electrophoresis, Nitrogen content 16.2%); was obtained from Siga- Aldrich Chemicals Ltd. (Germany).
- β-gal (from Aspergillus oryzae) was purchased from Siga- Aldrich Chemicals Ltd. (USA).
- Lactose (Pure Lab. Chemicals M.wt.360.31) was obtained from El-Nasr Pharmaceutical Co. for Chemicals, (Egypt).
- Sodium chloride (Purity 99.5%, M.wt.58.44) was obtained from Siga-Aldrich Chemicals Ltd. (Germany).
- Acetic acid (Purity 99.8%, M.wt.60.05) was obtained from Siga- Aldrich Chemicals Ltd. (Germany).
- Sodium acetate trihydrate (Purity 99%, M.wt.136.08) was obtained from El-Nasr Pharmaceutical Co. for Chemicals, (Egypt).
- Glucose kit (Enzymatic colorimetric method) was purchased from Diamond Diagnostics Co. for Modern Laboratory Chemicals, (Egypt).

- Total Protein kit (Colorimetric method) was purchased from Diamond Diagnostics Co. for Modern Laboratory Chemicals, (Egypt).

2.2.2 METHODS

2.2.2.1 MEMBRANE PREPARATION

Immobilized Cu^{+2} ions affinity cellophane –PGMA – grafted membranes have been prepared through three steps. The first step was introducing of epoxy groups to its chemical structure through grafting process with PGMA [9]. The second step was converting the introduced epoxy groups to sulfonic groups [9, 39].

2.2.2.2 IMMOBILIZATION OF CU^{+2} IONS

The third step; The cellophane sulfonated-grafted membrane was immersed in 20 mL of Copper sulfate solution (5 mM) in water bath for different temperatures (30:80°C) of pH (3:7) at (15:120 min). The amount of copper remaining in solution was then measured using atomic absorption spectrophotometer (Perkins- Elmer Analyst 300, USA) [40, 41].

2.2.2.3 PROTEIN ADSORPTION

The membrane immobilized with Cu^{+2} ions was immersed in protein solution, pH (2.6:5.2), containing β-Gal (0.005:1%) and BSA (0.05:1%) in a water bath at a temperature of 30°C for (1 to 5 h) [10].

2.2.2.4 PROTEIN ELUTION

The membrane immobilized with adsorbed Proteins was immersed in sodium chloride solution, pH (3:6), of different concentrations (0.1:3 M) in a water bath at temperature of 30°C for different time intervals (1 to 5 h) [10].

2.2.2.5 DETERMINATION OF β-GAL ENZYME AMOUNT

One milliliter of protein solution, adsorbing and/or eluting, was mixed with 100 mM lactose solution, pH 4.4, at 250 rpm for 30 min in room temperature. Samples were taken every 5 min to assess the glucose production using glucose kit. Enzyme activity is given by the angular coefficient of the linear plot of the glucose production as a function of time. β-Gal enzyme amount was determined as follows:

$$Amount\ of\ \beta - Galactosidase(mg) = \frac{\textbf{Measured protein solution activity}}{\textbf{Activity of 1mg of free enzyme}} \times 1\text{mg} \quad (1)$$

2.2.2.6 DETERMINATION OF BOVINE SERUM ALBUMIN AMOUNT

Using total protein kit, 20 μL of protein solution, adsorbing and/or eluting, was mixed with 1 mL reagent 2 (NaOH 0.2N, K-Na-tartrate 18 mM/L, potassium iodide 12 mM/L, cupric sulfate 6 mM/L) (A sample) and 20 μL of reagent 1 [protein standard 6 g/dL] with 1 mL reagent 2 (A standard). Mix and incubate for 5 min at 20–25°C. Measure the absorbance of sample (A sample) and standard (A standard) against reagent blank.

$$Bovine\ Serum\ Albumin\ amount\ (mg) = \frac{\textbf{A sample}}{\textbf{A standared}} \times 6 \quad (2)$$

2.2.2.7 MEMBRANE CHARACTERIZATION

2.2.2.7.1 WATER UPTAKE (WU %)

For water uptake measurement, membranes were immersed in distilled water at room temperature for 24 h and then their surfaces were dried by wiping with filter paper and weighing. The obtained results are the average of three samples [42–46].

$$W\% = \frac{Wt.of\ membrane\ (gm) - Wt.of\ dry\ membrane\ (gm)}{Wt.of\ Dry\ membrane(gm)} \times 100 \quad (3)$$

2.2.2.7.2 ENERGY DISPERSIVE ANALYSIS X-RAY

Elemental analysis of ungrafted and grafted membranes having different GPs was carried out using Energy Dispersive Analysis X-ray (Joel Jsm 6360LA, Japan).

2.3 RESULTS AND DISCUSSIONS

2.3.1 IMMOBILIZATION PROCESS

2.3.1.1 IMMOBILIZATION OF CU^{+2} IONS ON THE SULPHONATED GRAFTED CELLOPHANE MEMBRANES

To immobilize of Cu^{+2} ions (Fig. 2.1.) onto the sulphonated grafted cellophane membranes, the immobilization conditions have been studied and the obtained results are illustrated as follows.

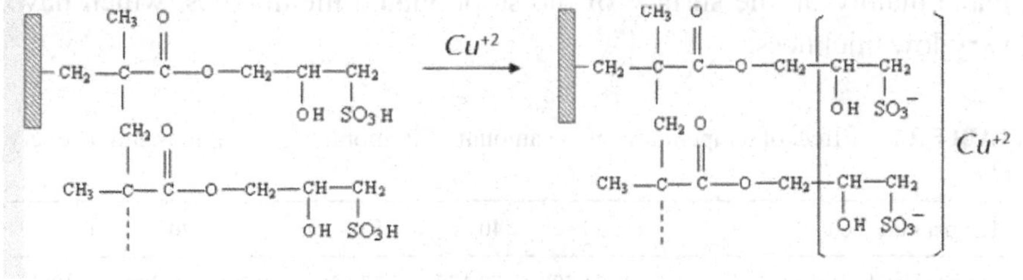

FIGURE 2.1 Mechanism of Immobilized Cu^{+2} ions on the sulphonated grafted cellophane membranes

2.3.1.2 EFFECT OF CU^{+2} IONS CONCENTRATION ON THE AMOUNT OF IMMOBILIZED METAL IONS AND RECOVERY (%)

The effect of variation Cu^{+2} ions concentration on the amount of immobilized metal ions given in Table 2.1. Show that no metal ions have been immobilized using Cu^{+2} concentration up to 0.25 mM. Further increase beyond this concentration, the Cu^{+2} ions uptake starts to increase almost linearly. The immobilized metal ions have been recovered totally from the membranes using diluted HCl solution.

TABLE 2.1 Effect of Cu^{+2} ions concentration on the amount of immobilized metal ions and recovery (%).

Concentration (mM)	0.10	0.25	0.50	1.00	2.00	3.00	5.00
Copper uptake (ppm/g)	0.00	0.00	5.58	14.95	39.62	41.59	68.90
Recovery (%)	0.00	0.00	99.88	98.35	99.44	94.74	98.60

2.3.1.3. EFFECT OF TEMPERATURE ON THE AMOUNT OF IMMOBILIZED METAL IONS AND RECOVERY (%)

No significant effect has been observed on the amount of immobilized Cu^{+2} ions upon increasing the temperature over 50°C; Table 2.2. This could be explained that the occurrence of the immobilization process takes place mainly on the surface of the sulphonated membranes, which have very low thickness.

TABLE 2.2 Effect of temperature on the amount of immobilized metal ions and recovery (%).

Temperature (°C)	30	40	50	60	70	80
Copper Uptake (ppm/g)	74.508	70.178	77.530	69.000	85.900	72.100
Recovery (%)	93.48	99.08	99.43	98.61	99.26	99.46

2.3.1.4 EFFECT OF THE TIME TAKEN FOR METAL IMMOBILIZATION'S TIME ON THE AMOUNT OF IMMOBILIZED METAL IONS AND RECOVERY (%)

Table 2.3 shows the effect of increasing immobilization time on the Cu^{+2} uptake over the membrane. Indeed no significant difference has been noticed with prolongation of the immobilization time from 15 min up to 120 min. These results confirm our conclusion from Table 2.2 (effect of temperature) that the immobilization process takes place mainly on the surface of the membranes, which have very low thickness.

TABLE 2.3 Effect of time on the amount of immobilized metal ions and recovery (%).

Time (min)	15	30	45	60	120
Copper uptake (ppm/g)	74.36	66.24	73.82	65.95	64.45
Recovery (%)	94.07	99.24	99.67	99.86	99.75

2.3.1.5 EFFECT OF METAL IMMOBILIZATION'S PH ON THE AMOUNT OF IMMOBILIZED METAL IONS AND RECOVERY (%)

The obtained data illustrated in Table 2.4 reveal the effect of pH on metal immobilization as determined from the amount of Cu^{+2} ions uptake. It can be seen that at pH 3, no Cu^{+2} ions have been immobilized. The metal ions started to be immobilized at pH 4.0 and increased with pH to reach the highest value at pH 5.0. Beyond this pH, the amount of Cu^{+2} ions uptake tends to level off. Generally, the recovery process of the immobilized Cu^{+2} ions has shown to be successful and all the immobilized amount of Cu^{+2} ions has been detected by the eluting solution.

TABLE 2.4 Effect of metal immobilized pH on the amount of immobilized metal ions and recovery (%).

pH	3	4	4.5	5	5.5	7
Copper Uptake (ppm/g)	0	50.33	43.48	64.25	61.31	64.89
Recovery (%)	0	100	100	100	100	100

2.3.1.6 EFFECT OF GRAFTING CONDITIONS ON THE AMOUNT OF SULPHONIC GROUP AND AMOUNT OF COPPER UPTAKE

The relationship between the metal uptake membranes' capability and the grafting conditions was evaluated. Figures 2.2–2.6 show the effect of the different grafting percentage on the character metal uptake. As the general trend shows that, increasing the grafting percentage is accompanied by decrease in the amount of Cu^{+2} ions uptake. The maximum amount of Cu^{+2} ions uptake (60 ppm/g) is obtained in case of membranes of the lowest grafting percentage (40%). On the other hand, for membranes with the highest grafting percentage (230%), the amount of uptake of Cu^{+2} ions is 32 ppm/g. Such observed trend might be explained by taking into account the effect of grafting conditions onto the obtained amount of sulphonic groups, which have been affected identically.

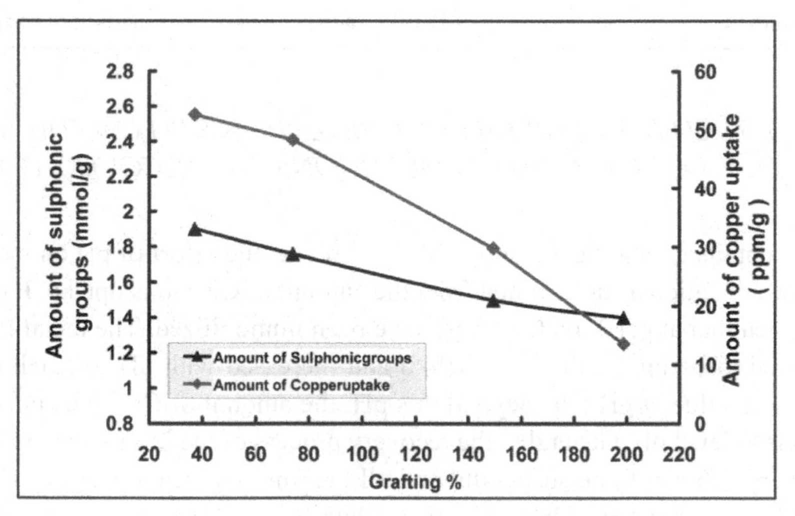

FIGURE 2.2 Effect of monomer concentration on the percentage of grafting, amount of sulphonic group and amount of copper uptake.

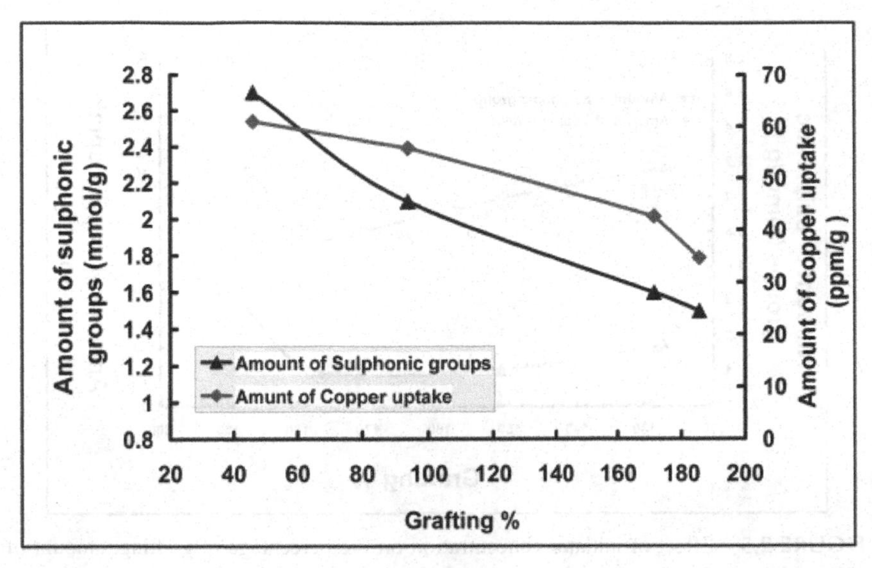

FIGURE 2.3 Effect of overall reaction temperature on the percentage of grafting, amount of sulphonic group and amount of copper uptake.

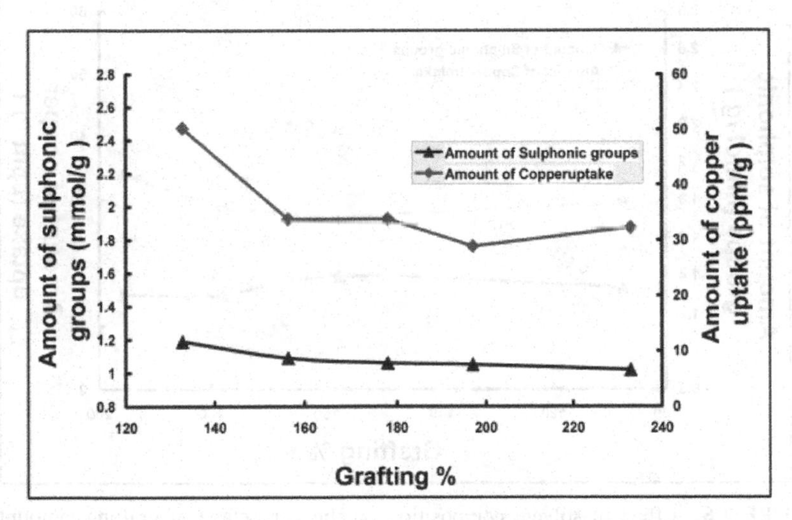

FIGURE 2.4 Effect of time on the percentage of grafting, amount of sulphonic group and amount of copper uptake.

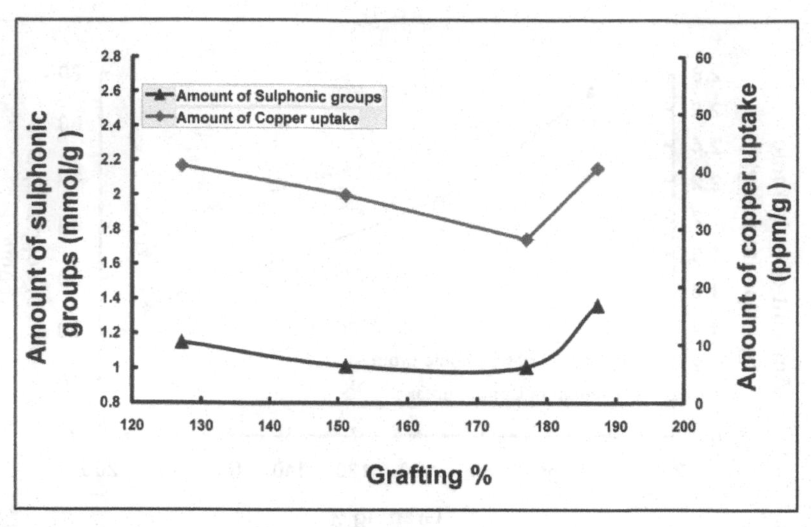

FIGURE 2.5 Effect of initiator concentration on the percentage of grafting, amount of sulphonic group and amount of copper uptake.

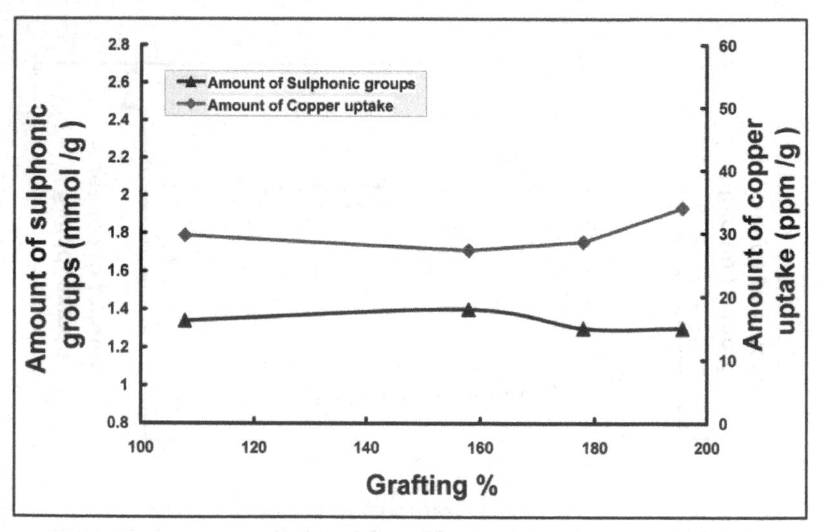

FIGURE 2.6 Effect of solvent composition on the percentage of grafting, amount of sulphonic group and amount of copper uptake.

2.3.2 MEMBRANE'S CHARACTERIZATIONS

2.3.2.1 WATER UP-TAKE

Water absorption in cellulose films is attributed mostly to the hydroxyl groups located on polysaccharides units. Nearly one-third of the cellulose in cellophane membranes is amorphous in nature and is responsible for water absorption, while crystalline zones being impenetrable. The data of water uptake of various grafted cellophane membranes together with that of ungrafted membranes is presented in Table 2.5. The data clearly indicate that the water uptake of the grafted cellophane membranes progressively decreased as the extent of grafting increased. This may be due to the fact that partial blockage of the internal membranes structure by the hydrophobic PGMA graft side chains. These results are in agreement with three obtained by the authors [47]. The effect of converting the epoxy groups, through sulphonation process, to sulphonated groups on the water uptake is presented also in Table 2.5. Almost the same trend has been observed in comparison with the grafted membranes in spite of the hydrophilic nature of the Sulphonated groups. The effect of the hydrophilic nature of sulphonated groups started to appear with membranes of grafting percentage higher than (70%). Immobilization of the sulphonated groups with Cu^{+2} ions implies different behavior of the water uptake. In general; the water uptake is higher than that of the sulphonated membranes. Maximum value was observed of membranes with grafting percentage from 22 to 35%. This behavior could be attributed to the amount of immobilized Cu^{+2} ions and its water chlechating nature.

TABLE 2.5 Water uptake of grafted, sulphonated and metal's immobilized cellophane membranes.

Gp (%)	9.53	22.14	35.23	70.53	92.79
Water uptake of Grafted Membranes, %	52.78	51.02	44.51	37.40	25.41
Water uptake of Sulphonated Membranes, %	51.21	53.50	45.11	40.25	35.05
Water uptake of Immobilized Cu^{+2}ions Membranes, %	54.42	59.76	60.65	44.34	42.52

2.3.2.2 SCANNING ELECTRON MICROSCOPY ANALYSIS

Increasing the grafting percentage increases the homogeneity of the obtained graft copolymers. In addition as preserved in Table 2.6, the thickness of the grafted membranes increased proportionally with grafting add-on. A concerning to the changes in the elemental structure as a result of the modification grafting process and subsequent sulphonation and metal immobilization process have been proved. The changes in C: O ratio as a result of the grafting process. The results tabulated in Table 2.7 indicate that C: O ratio changed in the favor of C as a result of PGMA add-on and the changes in S: Cu ratio as a result of the Cu^{+2} immobilization. Table 2.8 concludes such results, which illustrate that direct relation has detected between the amount of sulphonic groups and the immobilized Cu^{+2} ions.

TABLE 2.6 Effect of PGMA grafting on membrane thickness.

Gp (%)	0	135	275
Membrane Thickness (μm)	12	37	67.53

TABLE 2.7 Effect of PGMA grafting process on the C:O ratio.

Gp (%)	C	O
0	48.68	51.32
40.28	50.54	49.46
135	53.66	46.34

TABLE 2.8 Effect of PGMA grafting on the Cu^{+2} up-take and the S:Cu ratio.

	70.56	35.23	Gp, %
	1.76	1.9	Amount of Sulphonic Group (mM/g)
	69:31	65:35	S:Cu

2.3.2.3 PROTEINS SEPARATION

Membranes with the highest amount of immobilized Cu^{+2} ions were used in the study of the process of proteins separation (β-Gal and Bovine serum Albumin). IMAC membrane used in β-Gal separation step has been selected based on the following factors; Maximum immobilized Cu^{+2} ions, Minimum grafting percent and Maximum water-uptake. Different factors affecting both the proteins adsorption and elution steps have been evaluated and discussed in the light of membranes interaction with both proteins. The possible mechanisms for proteins adsorption onto the immobilized metal affinity membranes are categorized into four types of interactions [48–52]:

1. The binding affinity provided by the electron donating capacity of the imidazole groups of the exposed histidine residues on the protein surface with immobilized metal ions.
2. The electrostatic interaction between the charged protein molecules and positively charged metal ions.
3. The electrostatic interaction between the charged protein molecules and negatively charged sites on thc membrane surface such as the un-reacted hydroxyl groups from the basic cellulose materials and the sulphonation process or the residual SO_{3-} groups which did not check out with Cu^{+2} ions.
4. The hydrophobic interaction between the protein and the hydrophobic sites on the membrane surface; methyl groups from GA molecules.

The formed two interactions (1 and 2) contribute to specific binding between protein and immobilized metal, whereas the latter (3 and 4) two are nonspecific bindings. Table 2.9 reveals the non-specific interactions between contents in the mixture of proteins and the cellophane membranes in different stages (β-Gal and Bovine serum Albumin), which gave negative results nonspecific interactions between BSA and Cellophane membranes in any stages of membrane preparation. On the other hand, β-Gal shows very low hydrophobic interactions with p GA grafted membranes and low level of electrostatic interactions with sulphonated membranes. β-Gal adsorption with the three types of membranes in different ratio. This indicate that different of the molecules of BSA which are more bigger in size are compared with the molecules of β-Gal such difficultly of different molecules are non–specific between the three types of membrane with BSA.

TABLE 2.9 Non-specific interactions between BSA and β-Gal and cellophane, grafted cellophane and sulphonated grafted cellophane membranes.

β-Gal	BSA	Membrane Type
17%	Nil	Cellophane
5%	Nil	Cellophane-g-PGMA
15%	Nil	Cellophane-g-Sulphonated

2.3.2.4 PROTEINS ADSORPTION

The adsorption of proteins on the prepared membrane grafted, sulphonated and immobilized Cu^{+2} ions is affected by several factors such as Effect of β-Gal concentration, pH, Effect of adsorption's time and Effect of BSA: β-Gal ratio (Fig. 2.7.).

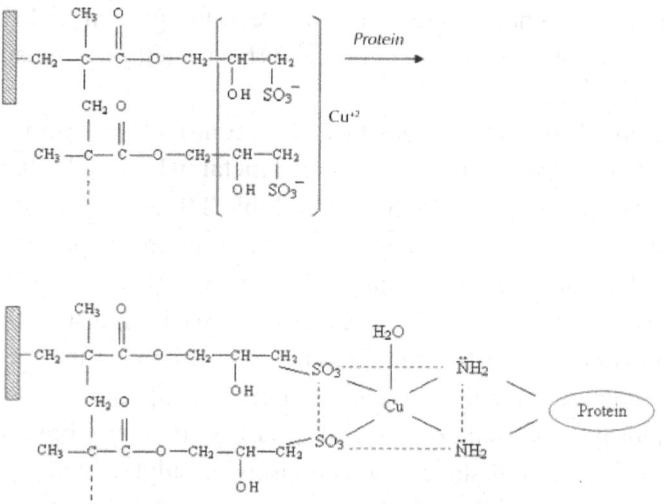

FIGURE 2.7 Mechanism of proteins adsorption on the immobilized Cu^{+2} ions on the sulphonated grafted cellophane membranes.

2.3.2.4.1 EFFECT OF β-GAL CONCENTRATION

The effect of variation of the concentration of β-Gal in the proteins mixture on the efficiency of membrane separation is shown in Fig. 2.8. The results reveal that increasing the β-Gal concentration in the proteins mixture increases the percentage of its adsorption onto the membrane. Maximum adsorption efficiency has been observed at 0.025% concentration. The adsorption isotherm is in agreement with the one obtained by other method (53). The presence of BSA doesn't affect the adsorption affinity of the membrane towards β-Gal although the initial concentration of β-Gal to BSA ranged from 1:100 up to 1:14. Since the protein adsorption behavior in the mixed protein system sometimes does not show the additive property of each protein, due to the interaction between them, the adsorption of β-Gal and BSA on the Cu^{+2} immobilized membrane in the mixed protein system was further investigated.

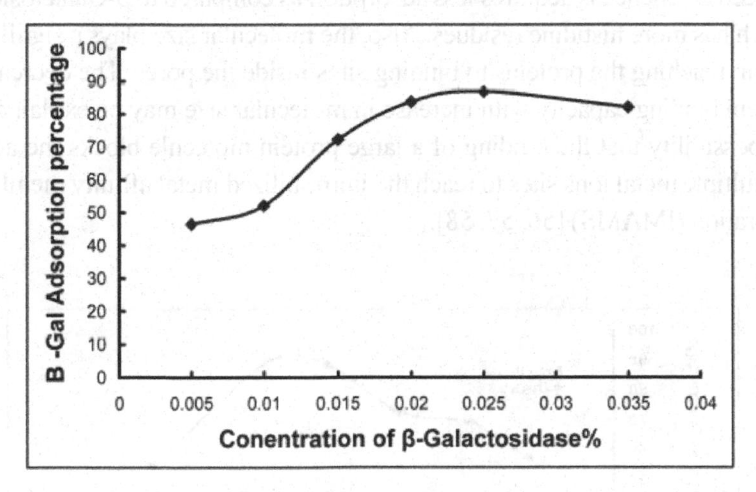

FIGURE 2.8 Effect of β-Gal. concentration in the proteins mixture on its adsorption percentage.

2.3.2.4.2 PH EFFECT

The percent adsorption results as a function of pH (from 2.6 to 5.2) presented in Fig. 2.9 show that BSA and β-Gal adsorption increased with increasing pH from 2.6 to 4.4 and then decreased slightly followed by decrease sharp respectively where the pH was raised further up to 5.2. The figure also reveals that the percentage of adsorbed β-Galactosidase is double as compared with that of BSA although the initial concentration of the latest is 20 times higher. Such this behavior could be explained in the light of the higher concentration of the incorporated amino acids located on the surface of β-Galactosidase as compared with the anther proteins. In addition, it is worthy to mention here that iso-electric point of both proteins is almost identical; PI= 4.7 for BSA and 4.61 for β-Galactosidase. This explains the reason of getting the higher percentage of proteins adsorption near this pH on which histidine, cystein, and tryptophan residues become neutral and consequently play an important role as a Ligand, which can coordinate to the immobilization Cu^{+2} ions [54]. In conclusion, the effect of pH on the protonation and hence the conformation changes of proteins has its reflection on the availability and number of coordinated histidine residues. It is known that BSA has only one or two to three exposed histidine [51, 55] and hence is acquires less adsorption as compared to β-Galactosidase, which has more histidine residues. Also, the molecular size plays a significant role in reaching the proteins to binding sites inside the pores. The decrease in protein binding capacity with increase in molecular size may be explained by the possibility that the binding of a large protein molecule blocks the access of multiple metal ions sites to reach the immobilized metal affinity membrane separation (IMAMS) [56, 57, 58].

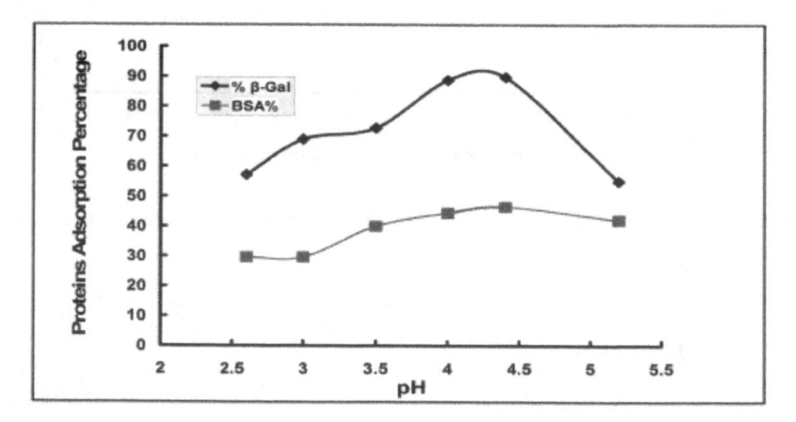

FIGURE 2.9 Effect of protein's solution pH on its adsorption percentage.

2.3.2.4.3 EFFECT OF ADSORPTION'S TIME

The effect of adsorption time on the Fig. 2.10, which has adsorption capacity of the IMACM towards both proteins is revealed furor seen that maximum adsorption capacity of the IMAC has been obtained after two hours. Further increase of adsorption is time, the adsorption capacity for both proteins decreased to reach equilibrium after three hours and four hours for β-Gal and BSA, respectively. The IMAC lost about 40% end 75% of its maximum adsorption capacity for β-Gal and BSA after reaching equilibrium. This behavior could be explained by the release of protein molecules tightly captured unstably immobilized copper ions on the IMAC. The difference in the percentage of released proteins may be explained by the difference in molecule size, which prevents the BSA from adsorption into the surface of the pores, so it became easier to release into the protein solution. On the other hand, β-Galactosidase small molecules have the chance to be absorbed into pores. Some of β-Gal molecules captured tightly the unstably immobilized Cu^{+2} ions and became harder to release into the protein solution from the pores. Only the part of β-Gal adsorbed onto the outer surface of IMAC is available to be released into the protein solution [59].

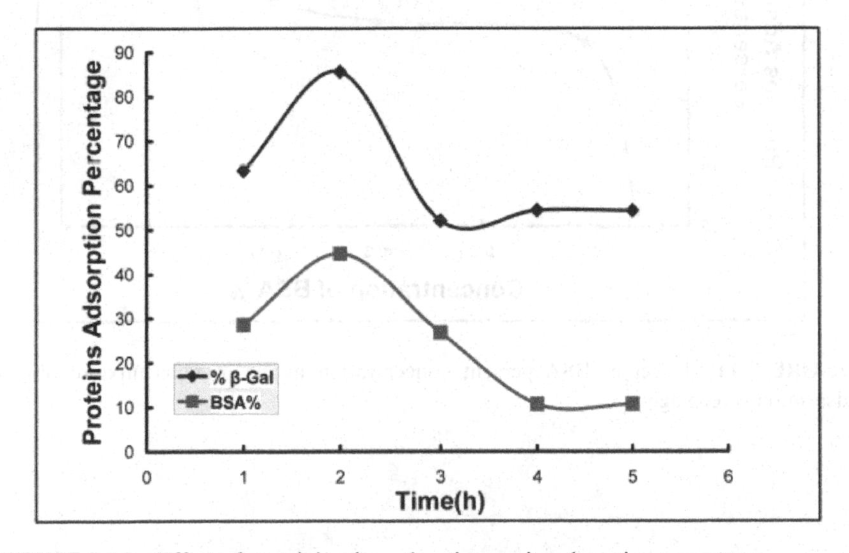

FIGURE 2.10 Effect of protein's adsorption time on its adsorption percentage.

2.3.2.4.4 EFFECT OF BSA: β-GAL RATIO

To give light on the affinity of the membranes towards β-Gal, solutions of proteins mixtures having different ratios starting from 1:1up to 1:20 have been used in the experimental adsorption of BSA. The results given in Fig. 2.11 show that at protein mixture of equal concentrations, each (0.025%), the results of IMAC show that the adsorbed percent of BSA is 0.0% while adsorbed percent of β-Gal is 90%. Increase the BSA concentration with keeping that of β-Gal constant leads to increase the adsorption of BSA gradually. This increment of BSA adsorption was not in favor of β-Gal since the adsorption of β-Gal from the mixed proteins solution mixture each of concentration 0.1% (are ratio of 1:1) is the same as that in case of 0.025% solution (β-Gal). This result reflects the affinity of IMAC towards β-Gal where its percent adsorption 90% in each case.

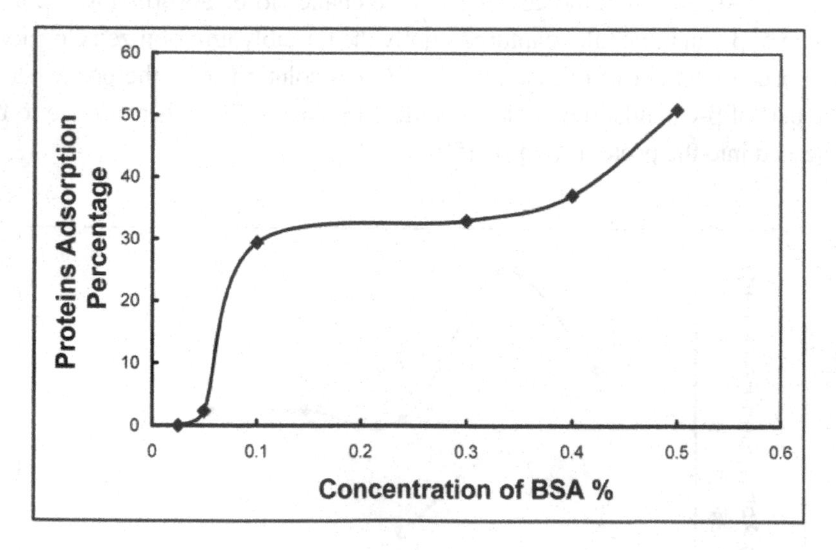

FIGURE 2.11 Effect of BSA percent concentration in the proteins mixture on its adsorption percentage.

2.3.2.5 PROTEIN ELUTION

In order to investigate the elution of both adsorbed proteins from IMAC, different factors affecting the elution process have been studied as following:

2.3.2.5.1 EFFECT OF ELUENT CONCENTRATION

Different concentrations of NaCl eluent have been used in the elution process of both β-Galactosidase and BSA proteins. Results given Fig. 2.12 shows that the percentage of eluted proteins increased with increasing of NaCl concentration. The maximum efficiency of proteins elution began to become apparent beginning from the NaCl concentration of 0.5 M till 2 M for the two molecules in the membrane for β-Galactosidase and BSA respectively. No effect has been observed on the eluted percentage in case of β-Galactosidase with increasing the concentration of eluent up to 3 M. In general, It is mentored [60] that NaCl solution in high concentration was effective for obtaining high proteins recovery, probably due to the shielding effect on proteins that are polyelectrolyte. It is essential to mention here that no Cu^{+2} ions leakage has observed when IMAC was treated under the same elution conditions.

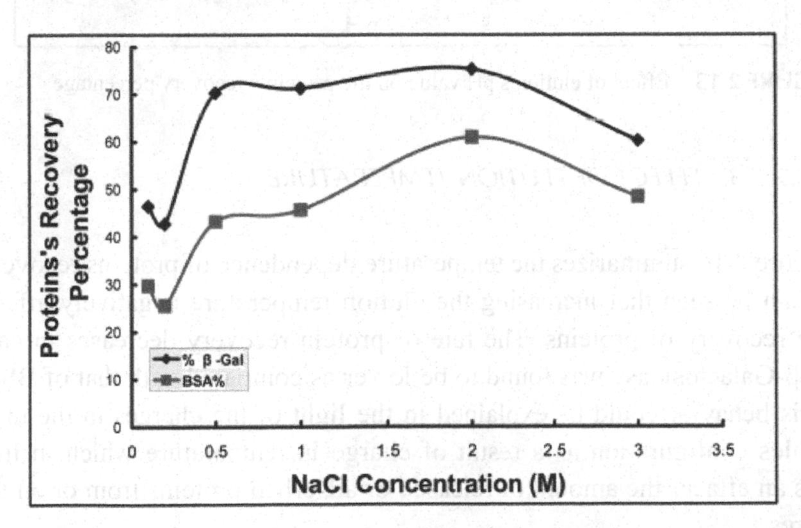

FIGURE 2.12 Effect of Eluent concentration (NaCl) on the proteins recovery percentage.

2.3.2.5.2 EFFECT OF PH

Figure 2.13 reveals the pH dependence of the recovered β-Galactosidase and BSA proteins when are different pH values was used 2 M NaCl solution as an eluent. For both proteins, the lower pH values of 4 and 3.5 for β-Gal and BSA, respectively, are effective to obtain the higher recovery. This finding can be explained by the pH dependence of the amounts of proteins adsorbed on IMAC [54].

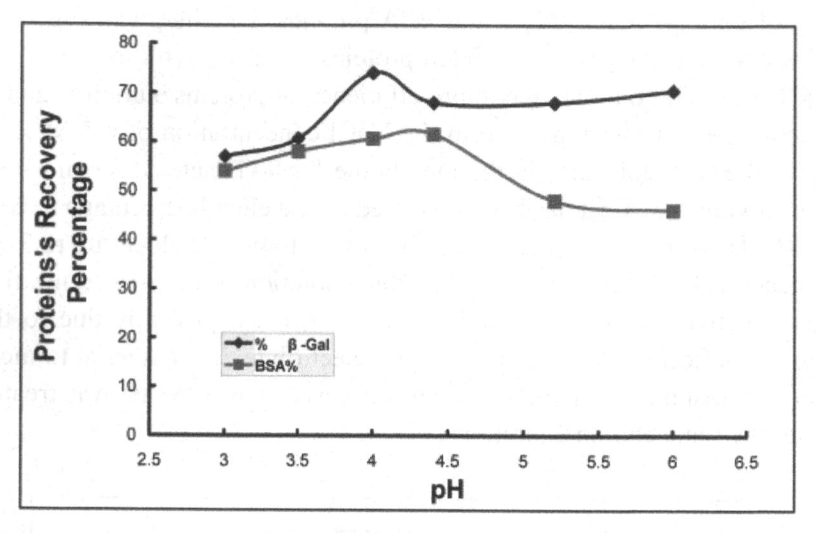

FIGURE 2.13 Effect of elution's pH value on the protein's recovery percentage.

2.3.2.5.3 EFFECT OF ELUTION TEMPERATURE

Figure 2.14 summarizes the temperature dependence of proteins recovery. It can be seen that increasing the elution temperature negatively affects the recovery of proteins. The rate of protein recovery decreases in case of β-Galactosidase was found to be lower as compared with that of BSA. This behavior could be explained in the light of the charges in the molecules configuration as a result of charge in temperature which in turn has an effacer the amount of released or adsorbed proteins from or on the pores.

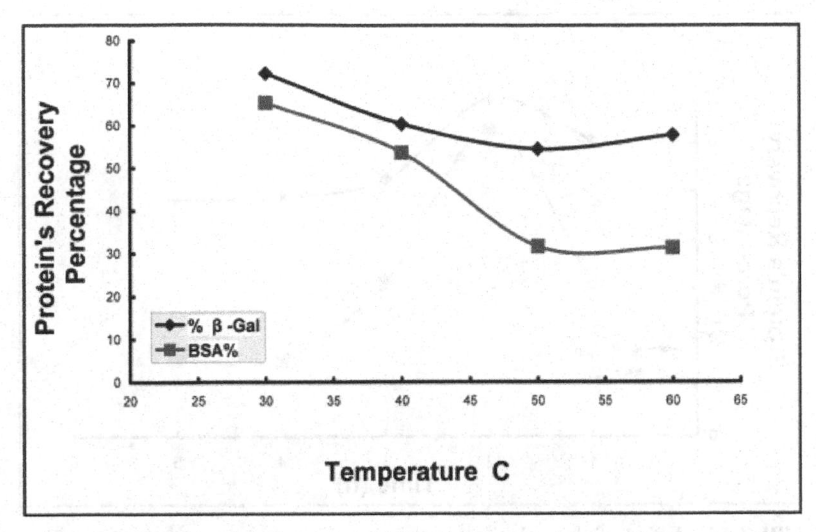

FIGURE 2.14 Effect of elution's temperature on the protein's recovery percentage.

2.3.2.5.4 EFFECT OF ELUTION TIME

The effect of elution's time on the efficiency of proteins recovery process is given in Fig. 2.15. Maximum recovery has been obtained after two hours of elution using 2 M NaCl at pH of 4.0 and 30°C. Prolongation of the elution time over than two hours is a companying by a decrease in the proteins recovery.

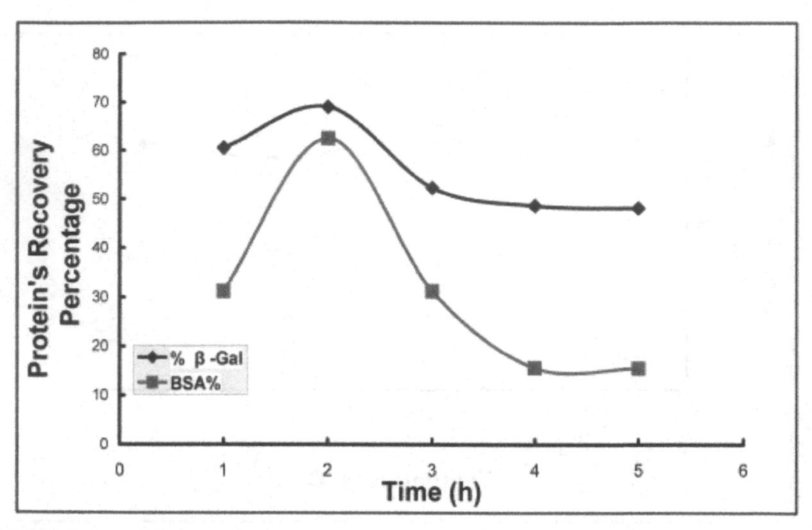

FIGURE 2.15 Effect of elution's time on the protein's recovery percentage.

2.4 CONCLUSIONS

The IMAC membranes showed high affinity toward the separation of β-Gal from its mixture with BSA, although the concentration of BSA, 0.5%, is 20 times higher than that of β-Gal concentration, 0.025%. Almost 90% of β-Gal has been adsorbed in comparison with 45% of BSA. This result implies the success of our technique in preparation of IMAC membranes for the separation of β-Gal enzyme from protein mixture.

KEYWORDS

- **enzymes**
- **graft polymers**
- **immobilized metal ion affinity chromatography**
- **membrane**
- **proteins**
- **separation techniques**

REFERENCES

1. Porath J., Carlsson J., Olsson I., Belfragge, G., Nature. 1975; 258: 598–599.
2. Fatiadi AJ., Crit Rev Anal Chem. 1987; 18: 1–44.
3. Porath, J., J Mol Recognit. 1990; 3: 123–124.
4. Porath, J., Protein Expr Purif. 1992; 3: 263–281.
6. Gaberc-Porekar V., Menart V., J Biochem Biophys Methods. 2001; 49: 335–360.
7. Mrabet NT, Vijayalakshmi MA., In: Vijayalakshmi MA, ed. Biochromatography: Theory and Practice. London: Taylor & Francis, Ltd. 2002; 272–294.
8. Ueda EKM, Gout PW, Morganti, L., J Chromatogr, A. 2003; 988: 1–23.
9. Mohy Eldin MS, Soliman EA, Hassan EA, Abu-Saied MA, J Appl Polym Sci.2009; 111: 2647–2656.
10. Abu-Saied MA (2006)" Preparation of Modified Natural Polymer for Its Utilization in Affinity Separation of Protein," M.Sc thesis, Al-Azhar University, Egypt.
11. Vasconcelos, H. L., Camargo, T. P., Goncalves, N. S., Neves, A., Laranjeira, M. M, C., Favere V. T., React. Funct. Polym.2008; 68, 572.
12. Wang, F., Guo, C., Liu, H. Z., C.-Liu, Z., J. Chem. Technol. Biotech. 2008; 83, 97.
13. Porath, J., Hansen, P., J. Chromatogr.1991; 550, 751.
14. Bayramoglu, G., Yakup Arica, M., Fibers and Polymers. 2012; 13, 225–1232
15. Bayramoglu, G., Kaya, B., Arica, M. Y., Chem. Eng. Sci.2002; 57, 2323.
16. Wang, L., Shen, S., He, X., Yun, J., Yao, K., Yao, S.-J., Biochem. Eng. J.2008; 42, 237.
17. Bayramoglu, G., Altntas, B., Arica, M. Y., Bioprocess. Biosyst. Eng.2011; 34, 127.
18. Gutierrez, R., Martin Del Valle, E. M., Galan, M. A., Sep. Purif. Rev.2007; 36, 71.
19. Wu, C. Y., Suen, S. Y., Chen, S. C., Tzeng, J. H., J. Chromatogr. A.2003; 996, 53.
20. Wu, F., Zhu, Y., Jia, Z., J. Chromatogr. A.2006; 1134, 45.
21. Oktem, H. A., Bayramoglu, G., Ozalp V. C., Arica, M. Y., Biotechnol. Progr.2007; 23, 146.
22. Dalal, S., Raghava, S., Gupta, M. N., Biochem. Eng. J.2008; 42, 301.
23. Lu, A. X., Liao, X. P., Zhou, R. Q., Shi, B., Colloid. Surf. A.2007; 30, 185.
24. Gaberc-Porekar V., Menart V., J. Biochem. Biophys. Meth. 2001; 49, 335.
25. Todorova Balvay, D., Pitiot, O., Bourhim, M., Srikrishnan, T., Vijayalakshmi, M., J. Chromatogr. B.2004; 808, 57.
26. Muszynska, G., Andersson, L., Porath, J., Biochem. 1986; 25, 6850.
27. Bayramoglu, G., Celik, G., M. Arica, Y., Colloid. Surf. A. 2006; 287, 75.
28. Bayramoglu, G., Yilmaz, M., M. Arica, Y., Biochem. Eng. J.2003; 13, 35.
29. Bayramoglu, G., Senkal, F. B., Celik, G., Arica, M. Y., Colloid. Surf. A. 2007; 294, 56.
30. Bayramoglu, G., Gursel, I., Yilmaz, M., Arica, M. Y., J. Chem. Technol. Biotechnol. 2012; 87, 530–539.
31. Xu, D., Hein, S., Wang, K., Mater. Sci. Technol.2008; 24, 1076.
32. Feng, Q., Wang, X. Q., Wei, A. F., Wei, Q. F., Hou, D. Y., Luo, W., Liu, X. H., Wang, Z. Q., Fiber. Polym.2011; 12, 1025.
33. Jin, G., Yao, Q. Z., Zhang, L., Chem. Res. Chinese. Univ.2008; 24, 154.
34. Feng, Z., Shao, Z., Yao, J., Chen, X., J. Biomed. Mater. Res. A. 2008; 86, 694.
35. Yang, C. L., Guan, Y. P., Xing, J. M., Liu, H. Z., Languir.2008; 24, 9006.

36. Arica, M. Y., Yilmaz, M., Yalcin, E., Bayramoglu, G., J. Chromatogr. B.2004; 805, 315.
37. Nie, H. L., T.-Chen, X., Zhu, L. M., Sep. Pur. Technol.2007; 57, 121.
38. Bayramoglu, G., Erdogan, H., Arica, M. Y., J. Appl. Polym. Sci.,2008; 108, 456.
39. Elkady MF, Abu-Saied MA, Abdel Rahman AM, Soliman EA, Elzatahry AA, Yossef ME, Mohy Eldin MS, Desalination. 2011; 279: 152
40. Kubota, N., Nakagawa, Y., Eguchi, Y, J Appl Polym Sci. 1996; 62, 1153.
41. Wu, C. Y., Suen, S. Y., Chen, S. C., Tzeng, J.H, J Chromatogr, A. 2003; 996, 53.
42. Mohy Eldin MS, Elzatahry AA, El-Khatib KM, Hassan EA, El-Sabbah MM, Abu-Saied MA, J Appl Polym Sci.2011; 119: 120
43. Abu-Saied MA, Elzatahry AA, El-Khatib KM, Hassan E A, El-Sabbah MM, Drioli E, Mohy Eldin MS, J Appl Polym Sci. 2012; 123: 3710
44. Mohy Eldin MS, Abu-Saied MA, Elzatahry AA, El-Khatib KM, Hassan EA, El-Sabbah MM, Int J Electrochem Sci.2011; 6: 5417
45. Abu-Saied, M. A., Fontananova, E., Drioli, E., Mohy Eldin, M. S., Journal of Polymer Research. 2013; 20: 187
46. Abu-Saied MA (2010) Preparation and Characterization of Polymer Polyelectrolyte Membranes for Direct Methanol Fuel Cell Application, PhD thesis, Al-Azhar University, Egypt.
47. El-Awady Nagwa, I., El-Awady, M. M., El-Din, M. S.Mohy, J.Text.Polym. Sci. Technol. 1999; 3,25: 41.
48. Beeskow, T. C., Kusharyoto, W., Anspach, F. B.,. Kroner, K.H, W. Deckwer, D., J. Chromatogr. A.1995; 49,715.
49. Tsai, Y. H., Wang, M. Y., Suen, S. Y., J. Chromatogr. B. 2002; 133, 766.
50. Porath, J., Olin, B., Biochemistry.1983; 22, 1621.
51. Sharma, S., Agarwal, G. P., Anal. Biochem.2001; 126,288.
52. Chen, W. Y., Wu, C. F., Liu, C. C., J. Colloid Interface Sci.1996; 135,180.
53. Xianfang Zeng, Eli Ruckenstein, J. Membr. Sci.1999; 157; 97: 107.
54. Naoji Kubota, Yasuhiro Nakagawa, Yukari Eguchi, J.Appl.Polym.Sci.1996; 62: 11153–1160.
55. Gaberc-Porekar V., Menart V., J. Biochem. Biophys. Methods publication).2001; 49, 335.
56. Camperi, S. A., Grasselli, M., A. A. Navarro del Canizo, Smolko, E. E., Cascone, O., J. Liq. Chromatogr. Relat. Technol. 1998; 21,1283.
57. Yang, L., Jia, L., Zou, H., Zhou, D., Zhang, Y., Sci. China, Ser.B: Chem.1998; 41, 596.
58. Crawford, J., Ramakrishnan, S., Periera, P., Gardner, S., Coleman, M., Beitle, R., Sep. Sci. Technol. 1999; 34, 2793.
59. Chun-Yi Wu, Shing-Yi Suen, Shiow-Ching Chen, Jau-Hwan Tzeng, J. Chromatogr. A. 2003; 996, 53: 70
60. Naoji Kubota, Yasuhiro Nakagawa, Yukari Eguchi, J.Appl.Polym.Sci.1996; 62: 11153–1160.

CHAPTER 3

SATELLITE IMAGING FOR ASSESSING THE ANNUAL VARIATION OF FISH CATCH IN EAST AND WEST COAST OF INDIA

C. O. MOHAN, B. MEENAKUMARI, A. K. MISHRA, D. MITRA, and T. K. SRINIVASA GOPAL

CONTENTS

3.1 Introduction 138
3.2 Materials and Methods 140
3.3 Results and Discussion 142
3.4 Conclusion 157
Acknowledgments 158
Keywords 158
References 158

3.1 INTRODUCTION

Identifying the potential fishing zones (PFZ) is very useful as it minimizes the time and fuel spent on searching the fish resources. In the present study, the annual chlorophyll *a* and sea surface temperature variations in east and west coast of India was investigated using Moderate Resolution Imaging Spectroradiometer (MODIS) ocean color data and it was correlated with the fish catch data. The results of the study indicated that the chl *a* concentration ranged from 0.1 to 60 mg m^{-3} and SST from 17 to 28°C in both the coasts. Chlorophyll *a* concentration was twice higher in west coast as compared to east coast throughout the study period. There was a positive relation between the total fish catch and chl *a* in both the coasts, which can be used for predicting the fish catch thereby reducing the fuel spent on searching fish shoal.

Marine capture fisheries are the most diverse of the major global food-producing sectors, both in terms of the range of species harvested [1] and harvesting technologies used. One characteristic, which is common to world fisheries, is their dependence on fossil fuels. The research indicates that direct fuel inputs typically account for between 75 and 90% of total energy inputs to fishing activities [2–4]. The scale of direct fuel inputs, however, can range widely. Purse seine fisheries for small pelagic species typically use under 50 L of fuel per ton of fish landed [4, 5], whereas the fisheries targeting high value species frequently consume in excess of 2000 L per ton of landings [5–7]. In 2000, world's fisheries fleets burned approximately 50 billion L of fuel, representing approximately 1.2% of total global oil consumption [8], for catching 80.4 million t of fish and invertebrates from marine waters [9], releasing approximately 134 million t of CO_2 into the atmosphere. This accounts for the global average fuel use intensity of 620 L per live weight tons of fish and shellfish landed [8]. Major portion of the fuel consumption is attributed to locate the fisheries resources. Fisheries resources are not distributed uniformly and their abundance varies with the ocean features such as water color, turbidity, sea state, flotsam and jetsam, wave size and direction, wind patterns and temperature distribution over the sea surface. Fluctuation in any of the environmental conditions over the ocean affects the distribution, abundance and availability of fish. Therefore, there is a need for a systematic scientific observation and detailed study of the ocean both spatially and temporally. This is achieved by adopting the satellite remote sensing technology.

Sea surface temperature (SST) can be used for monitoring the fish distribution as fishes prefer certain temperature within the habitat [10] and also as temperature impacts directly on all life-stages of fish. Apart from SST, phytoplankton also plays a major role in the fish distribution. Phytoplankton are floating or drifting single-cell algae that are primarily transported by water motion [11, 12]. These are found in all estuarine waters and in open oceans to a lesser extent and contribute greatly to overall primary production. Phytoplankton contains chloroplasts, which absorb and use the underwater light to fix carbon in the form of carbohydrate. Among the chloroplast pigments, chlorophyll-*a* is common to all phytoplankton [13]. Thus, chlorophyll-*a* is an indicator of the abundance of phytoplankton in the water. As the primary producers form the base of the aquatic food web, monitoring primary productivity via chlorophyll *a* concentrations is a useful tool in fisheries resource management [14]. The pronounced scattering/absorption features of chlorophyll-*a* are: strong absorption between 400–500 nm (blue) and at 680 nm (red), and reflectance maximums at 550 nm (green) and 700 nm (near-infrared (NIR) [15]. Use of satellite image data to investigate oceanic processes has become an essential component of oceanographic research and monitoring. Data from the Coastal Zone Color Scanner (CZCS) provided the first demonstration of the ability to observe the abundance and distribution of phytoplankton chlorophyll in the world's ocean from space [16–18]. Over the years, many satellites are being used to monitor the ocean color and SST. The Moderate Resolution Imaging Spectroradiometer (MODIS) is an environmental satellite operating in visible, near- and short-wave infrared and thermal portions of the electromagnetic spectrum and acquires images in 36 spectral bands observing the land, atmosphere, and oceans of the Earth between 412 to 14,385 nm [19]. It has spatial resolution of 250, 500 and 1000 m depending on the spectral band and has a swath width of 2330 km, enabling the entire surface of the Earth to be viewed every 2 days. MODIS images have the potential to provide quantitative measures of numerous geophysical parameters, including chlorophyll *a* and SST.

India, which is positioned in the central part of the Indian Ocean, with a long coastline of 8118 km, contributes significantly to the marine fish production [20]. Its marine resources are spread over in the Indian Ocean, Arabian Sea, and Bay of Bengal. The exclusive economic zone (EEZ) of the country has an area of 2.02 million km^2 comprising 0.86 million km^2 on the west coast, 0.56 million km^2 on the east coast and 0.6 million km^2 around the Andaman and Nicobar islands. Estimates of potential fishery resources from the EEZ of India are about 3.5 to 4.7 mt (million tons) [21–23]. The recent estimates on annual marine landings from the Indian coast show that they fluctuate between 2.2 and 3.0 million tons [24]. Of this, about 73% of the catches originate from the west coast of India. The Arabian Sea, which is situated in the west coast of India, is one of the most productive oceanic areas in the world [25, 26]. The Bay of Bengal in the eastern part of the north Indian Ocean is a tropical basin. Efforts have been made to study the chlorophyll distributions in the Arabian Sea [27–29], Bay of Bengal [30–33] and its influence on the sea surface temperature [34], whereas there is very limited effort has been made to correlate these parameters with the actual fish catch. Hence the present study was undertaken with the objective of evaluating the correlation of the chlorophyll *a* and sea surface temperature derived from the satellite MODIS with the fish catch from 2002 to 2006.

3.2 MATERIALS AND METHODS

3.2.1 STUDY AREA

The study area consists of Bay of Bengal in east and Arabian Sea in west coast of India (Fig. 3.1). In west coast, the study area extended from 7° N to 24° N latitude and 65° E to 77° E longitude, whereas in east coast it extended from 7° N to 21° N latitude and 77° E to 91° E longitude.

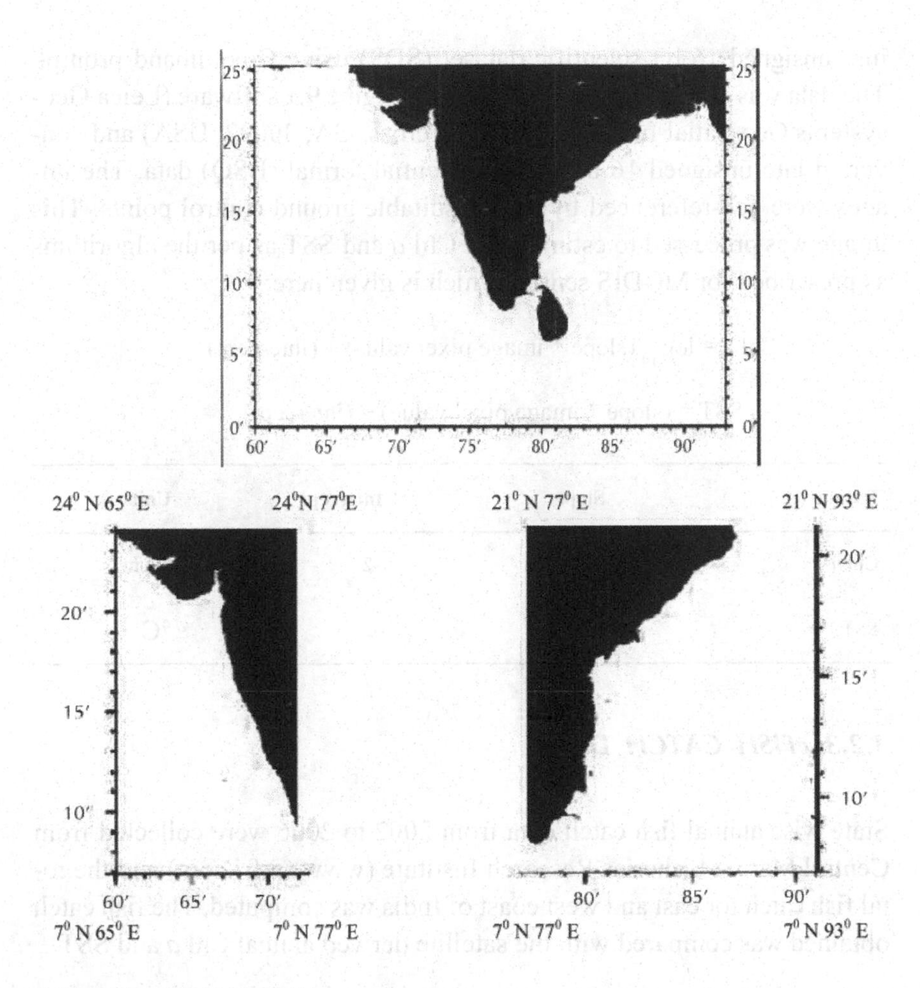

FIGURE 3.1 STUDY AREA.

3.2.2 SATELLITE DATA PROCESSING

The annual composite level II data (2002 to 2006) of chlorophyll *a* (Chl *a*) and sea surface temperature (SST) from MODIS sensor were obtained from GSFC (Goddard Space Flight Center), NASA, USA (http://oceancolor.gsfc.nasa.gov) in hierarchial data format (HDF). The data was converted into unsigned 16 bit scientific dataset (SDS) using C-command prompt. The data was imported into the ERDAS Imagine 9.1 software (Leica Geosystems Geospatial Imaging, LLC, Norcross, GA, 30092, USA) and converted into unsigned 16 bit band sequential format (BSQ) data. The images were geo-referenced by taking suitable ground control points. This image was processed to estimate the Chl *a* and SST as per the algorithms as prescribed for MODIS sensor, which is given here:

$$\text{Chl a} = \log_{10} (\text{slope} * \text{image pixel value}) + (\text{intercept})$$

$$\text{SST} = (\text{slope} * \text{image pixel value}) + (\text{intercept})$$

Parameter	**Slope**	**Intercept**	**Units**
Chl *a*	0.015	–2	mg m^{-3}
SST	0.05	–2	°C

3.2.3 FISH CATCH DATA

State wise annual fish catch data from 2002 to 2006 were collected from Central Marine Fisheries Research Institute (www.cmfri.com) and the total fish catch for east and west coast of India was computed. The fish catch obtained was compared with the satellite derived annual Chl *a* and SST.

3.3 RESULTS AND DISCUSSION

3.3.1 ANNUAL CHLOROPHYLL A VARIATION

The surface Chlorophyll *a* concentration in east and west coast of India were found to be different (Figs. 3.2a–3.6a). The concentration of the chlorophyll *a* ranged from 0.1 to 60.2 mg m^{-3} (Table 3.1. In east coast, the chl *a* concentration over the years ranged from 0.1 to 50.6 mg m^{-3}, with mean concentration ranging from 0.25 to 0.31 mg m^{-3}. The concentration was higher (>3 mg m^{-3}) (yellow color), in the extreme north-eastern coast and south-eastern coast compared to other areas in the east coast and a concentration >8 mg m^{-3} (red color) was present only in certain pockets of the east coast along the continental shelf area. Dey and Singh [32] have also reported a concentration of >2 mg m^{-3} along the south-eastern region. The chlorophyll *a* values obtained in the present study were in agreement with the in-situ values reported by Sharma et al. [35] along the east coast of India. In west coast, Chl *a* concentration was in the range of 0.1 to 60.2 mg m^{-3}, with a mean of 0.51 to 0.6 mg m^{-3}, which is twice as that of east coast. A very high concentration (>8 mg m^{-3}) was observed prominently in north-western part along the continental shelf area. Authors [32] have also reported similar chlorophyll concentrations in the north-western coast of India using IRS-P4 and OCTS images, respectively. A concentration of over 3 mg m^{-3} (yellow color) was observed all along the west coast, however, with slightly broader area in the north-western region. The contour of 0.6–1.0 mg m-^{3} concentration (green color) runs almost parallel to the coast in both the coasts. However, its spread was narrower in east coast, except in the south-eastern region, whereas in the west coast, it showed a very broad area under this concentration. A consistent decrease in chlorophyll from the coast to the offshore is clearly seen and the chlorophyll *a* concentration showed very little variations, both spatially and temporally (<0.3 mg m^{-3}) in both the coasts. Similar results were reported by Sharma et al. [35] in east coast of India.

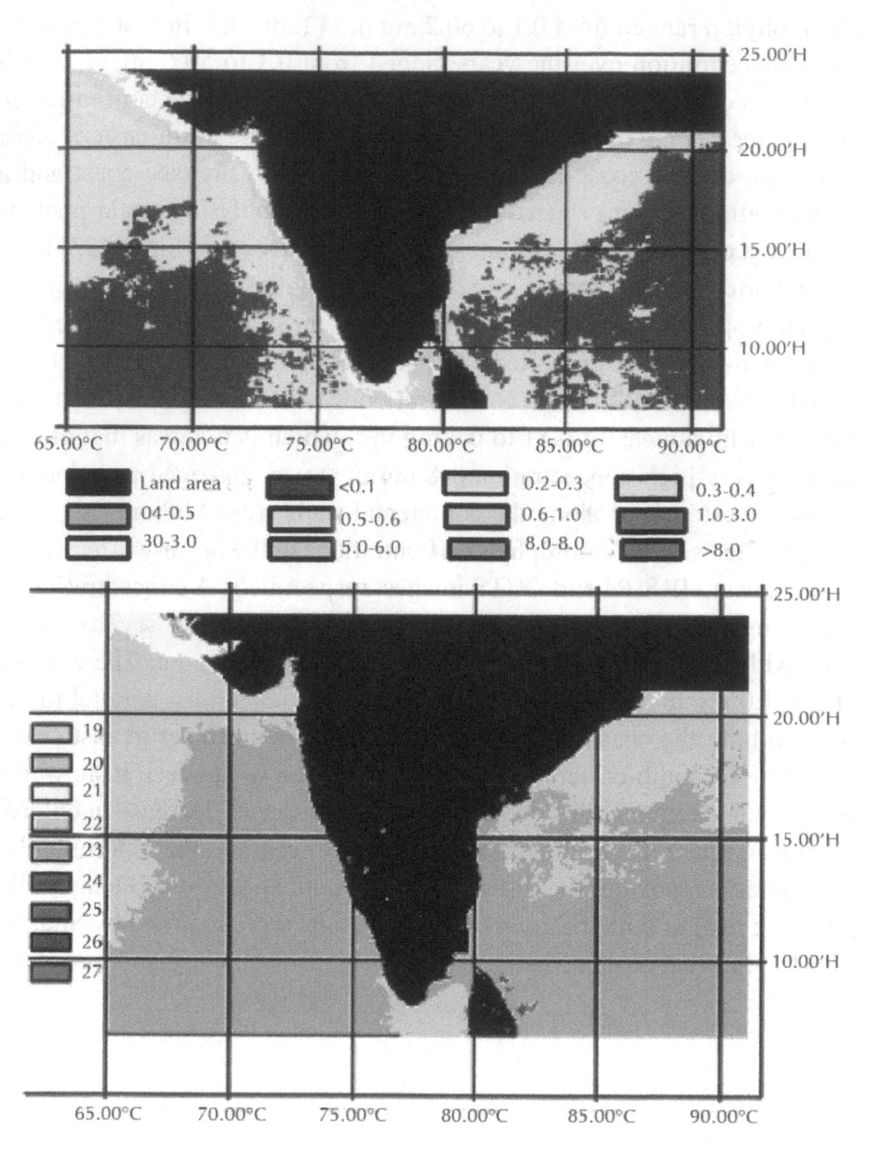

FIGURE 3.2 Chlorophyll *a* concentration (a) and SST (b) derived from MODIS for 2002.

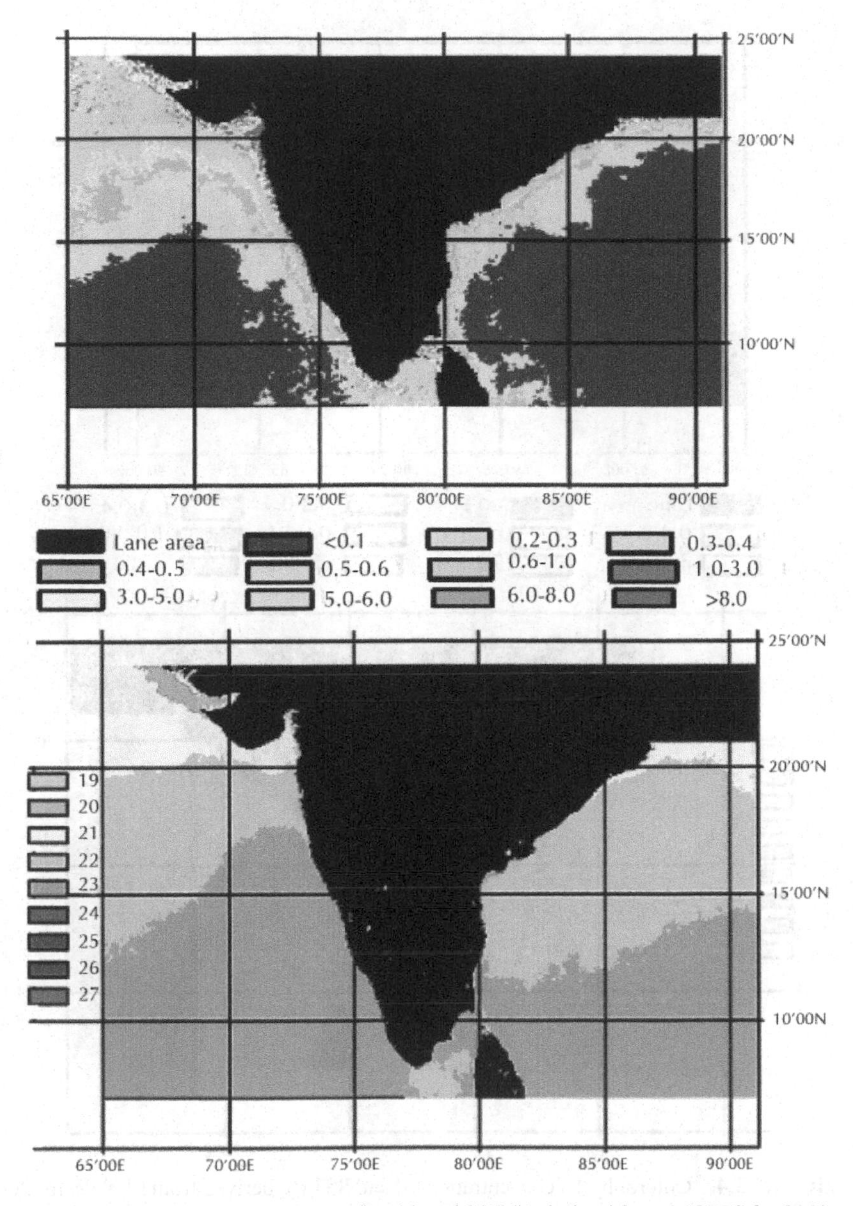

FIGURE 3.3 Chlorophyll *a* concentration (a) and SST (b) derived from MODIS for 2003.

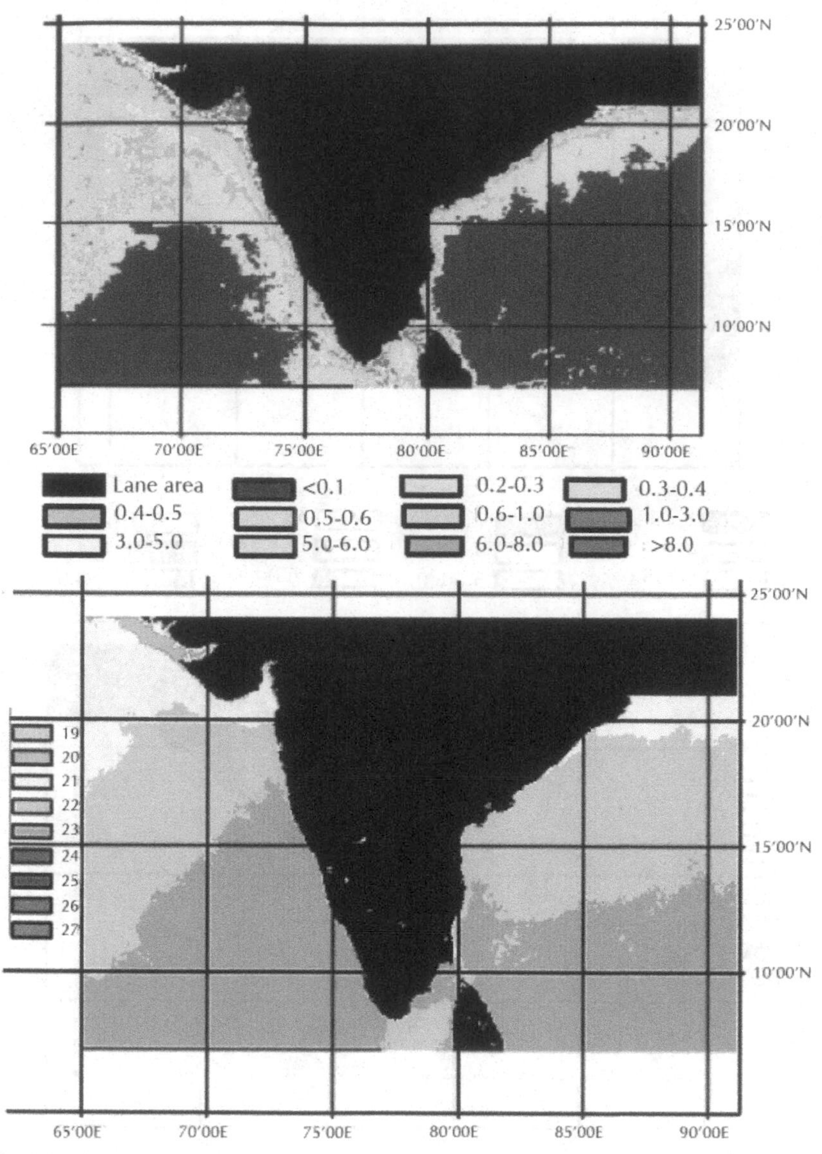

FIGURE 3.4 Chlorophyll *a* concentration (a) and SST (b) derived from MODIS for 2004.

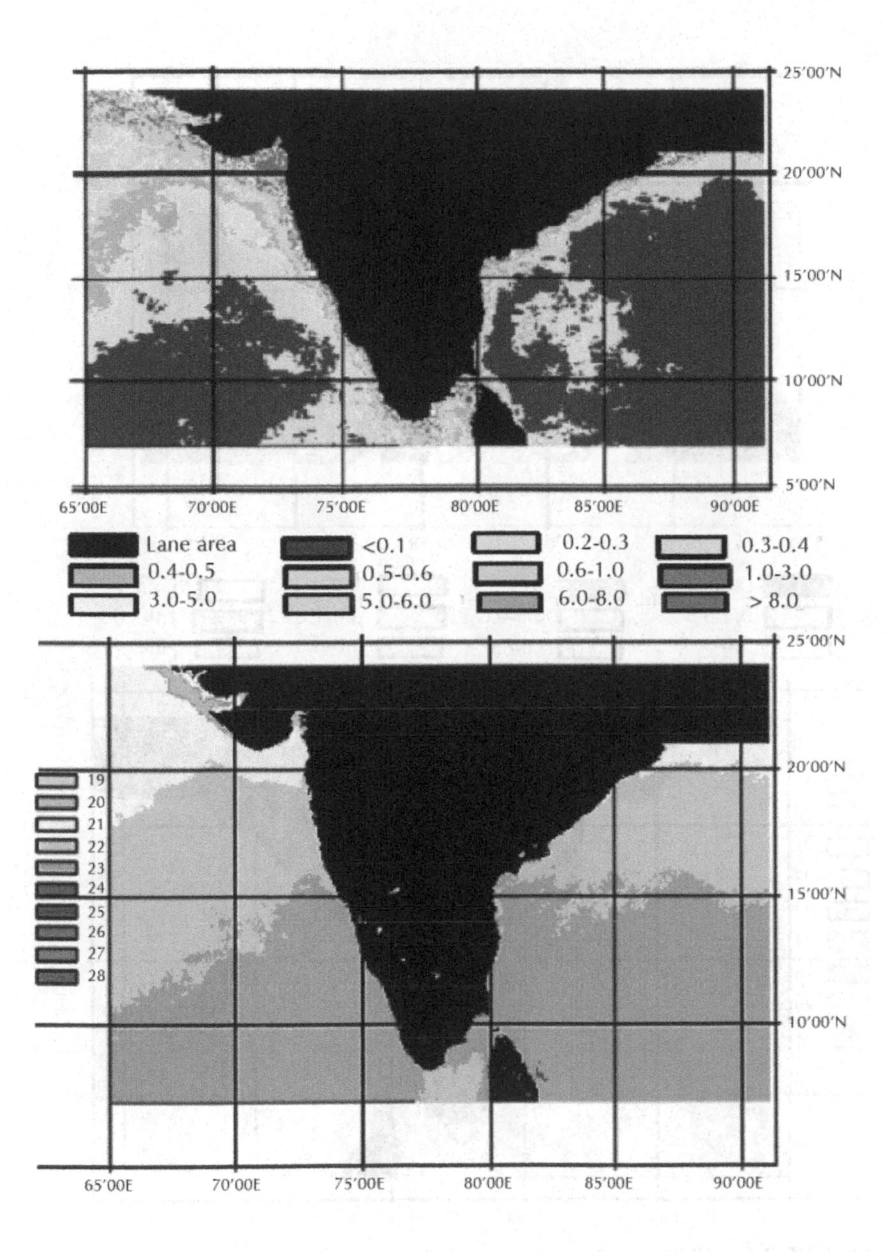

FIGURE 3.5 Chlorophyll *a* concentration (a) and SST (b) derived from MODIS for 2005.

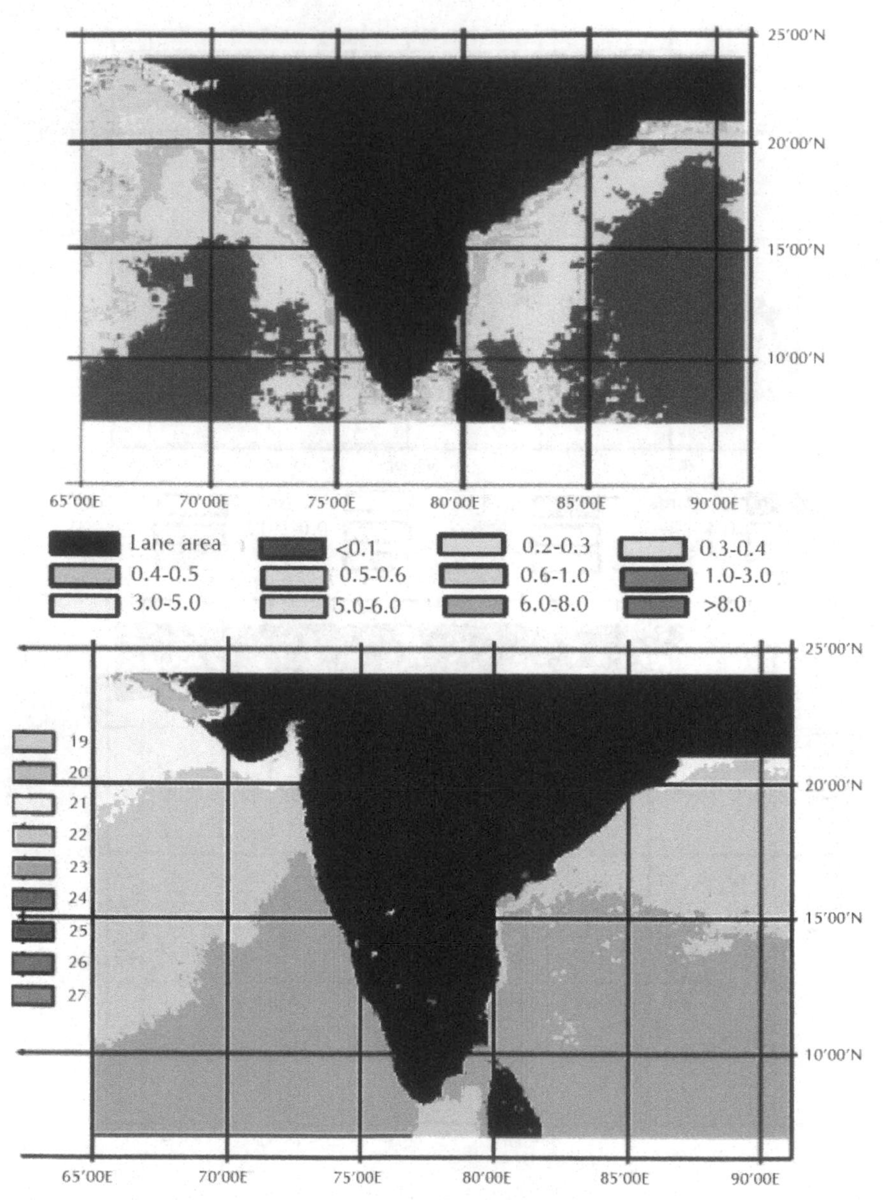

FIGURE 3.6 Chlorophyll *a* concentration (a) and SST (b) derived from MODIS for 2006.

TABLE 3.1 Range of SST and Chl *a* concentration of east and west coast of India during 2002–2006.

Year	Sea surface temperature (°C)		Chlorophyll *a* (mg m^{-3})	
	East coast	West coast	East coast	West coast
2002	21–28	19–27	0.1–50.6	0.1–20.6
2003	20–27	19–27	0.1–25.4	0.1–22.1
2004	20–27	19–27	0.1–13.6	0.1–19.9
2005	17–28	20–26	0.1–14.1	0.1–60.2
2006	20–27	20–27	0.1–15.1	0.1–52.4

The high values of chlorophyll concentration in the extreme south-eastern and north-eastern regions compared to the other parts of the Bay of Bengal is attributed to the very low depth (10–15 m) of the region [32]. This leads to the vertical mixing of the saline water with the fresh water carried by the rivers in this region and upwelling of the nutrients. In other regions of the Bay of Bengal, the vertical mixing and the upwelling of the nutrients are not strong enough to support high primary production. Traditionally, the east coast of India, which is guarded by the Bay of Bengal is considered as a low biological production region, stems primarily from the lack of availability of nutrients in the upper layers arising from strong stratification and weaker winds and excessive freshening. The open ocean water in the Bay of Bengal has lower nutrient concentrations compared to those in the Arabian Sea. The low nutrient content in the ocean water of the Bay of Bengal is the prime cause of the low chlorophyll concentration in the Bay of Bengal. This has been variedly attributed to the light inhibition due to turbidity and/or cloud cover, narrow continental shelf. Low nutrient content is also attributed to the low nitrate and silicate content of the river runoff water and its consumption within the estuarine and coastal region [31]. The other reason for the low nutrient content is the upwelling, which is confined very close to the coast (mostly within 40 km) along the south-western boundary and is periodic. One of the reasons for the lack of intense upwelling along the western boundary of the Bay of Bengal during summer, in spite of the upwelling-favorable south-westerly winds, is the equator-ward flow of the freshwater plume, which could overwhelm the offshore Ekman transport. The freshening is in part due to oceanic precipitation as well as by the run off from peninsular rivers such as Brahmaputra, Ganges, Irrawady, Godavari, Mahanadi, Krishna, and Kaveri. Abundant rainfall and river water freshens the upper layers of the Bay of Bengal even during summer due to the perennial river runoff leads to a strongly stratified surface layer as compared to Arabian Sea. The weaker winds over the Bay are unable to erode the strongly stratified surface layer, thereby restricting the turbulent wind-driven vertical mixing to a shallow depth of < 20 m [31]. In contrast, the productivity of the Arabian Sea is brought about through a range of physical processes. The coastal upwelling along Somalia, Arabia and southern part of the west coast of India turns the coastal waters into a region of high biological productivity. The open ocean upwelling, wind-driven mixing and lateral advection [26] makes the open ocean waters of the central Arabian Sea more productive.

3.3.2 ANNUAL SEA SURFACE TEMPERATURE VARIATION

Temperature is an important parameter, which has a profound influence on all the life forms including phytoplanktons and fishes. In the present study, the sea surface temperature in both the coasts varied from 17 to 28°C (Table 3.1), however, the temperature of east coast showed 1–2°C higher compared to Arabian Sea. The results obtained in the present study are consistent with [31]. The temperature in the north-western coast was less than 20°C throughout the study period (Figs. 3.2b–3.6b) compared to other regions. In general, the temperature in the north-western, north-eastern and extreme south-eastern regions was less than the south-western and north-eastern regions. Temperature in the coastal water was higher than the oceanic waters. The mean temperature of both the coasts was around 22.3 to 22.7°C in both the coast, with slightly lower temperature in the west coast. This relatively lower SST in Arabian Sea could be attributed to the prominent upwelling and convective mixing in the west coast, which brings the denser, cool water into the sea surface from the thermocline regions. The higher SST in the Bay of Bengal could be due to the presence of strong stratified surface layer, which restricts the vertical mixing [31].

3.3.3 ANNUAL FISH CATCH VARIATION AND ITS CORRELATION WITH CHL A AND SST

Annual fish catch of east and west coast varied greatly during 2002 to 2006. In east coast, the total fish landings ranged from 7.5 to 8.8 lakh tons compared to 15.4 to 18.4 lakh tons in the west coast (Tables 3.2 and 3.3). In east coast, the fish catch was minimum during 2005 (7.5 lakh tones) and maximum during 2004, at which the chl *a* concentration was 0.23 and 0.31 mg m^{-3}, respectively. Similar to east coast, minimum fish catch was observed during 2005 (15.4 lakh tons) in west coast also at which chl *a* concentration was 0.54 mg m^{-3}. In west coast, the maximum fish catch was observed during 2006 (18.4 lakh tones) with annual chl *a* concentration of 0.59 mg m^{-3}. In general, the results indicated that as the chl *a* concentration increases, fish catch increases in both the coasts (Fig. 3.7). Positive correlation was obtained between fish catch and chl *a* concentration in both the coast (Fig. 3.8). However the correlation coefficient (r^2) was higher for west coast (0.63) than the east coast (0.42). Although the landings of both the coasts varied greatly, the mean SST was around 22°C throughout the study period (Tables 3.2 and 3.3). In east coast, the mean SST was 22.6 and 22.4°C during the year of lowest and highest fish landings, where as in west coast it was 22.28 and 22.34°C, respectively. In east coast, the fish catch increased with the decrease in the SST, where as this relation was not observed prominently in west coast (Fig. 3.9). The negative correlation was observed between the fish catch and SST in east coast compared to west coast (Fig. 3.10). However the correlation coefficient was insignificant in both the coasts. Although the correlation coefficient was insignificant, a positive correlation was observed between the SST and chl *a* concentration in both the coasts (Fig. 3.11). In general, the annual fish catch was 65 to 68.7% in west coast and 31.2 to 34.9% in east coast of India. This could be attributed to the high chl *a* concentration, thereby making the region highly productive.

TABLE 3.2 Mean SST, Chl *a* and fish catch of east coast of India during 2002–2006.

Year	Sea surface temperature (°C)	Chlorophyll *a* (mg m^{-3})	Total catch (*t*)	% of total catch
2002	22.71 ± 0.48	0.299 ± 0.68	809999	31.28
2003	22.45 ± 0.57	0.249 ± 0.57	824638	31.87
2004	22.42 ± 0.6	0.315 ± 0.55	885609	34.89
2005	22.62 ± 0.61	0.237 ± 0.89	749385	32.64
2006	22.65 ± 0.51	0.265 ± 0.49	866749	31.97

TABLE 3.3 Mean SST, Chl *a* and fish catch of west coast of India during 2002–2006.

Year	Sea surface temperature (°C)	Chlorophyll *a* (mg m^{-3})	Total catch (*t*)	% of total catch
2002	22.65 ± 0.57	0.588 ± 0.14	1779646	68.72
2003	22.47 ± 0.77	0.567 ± 0.12	1762457	68.13
2004	22.34 ± 0.79	0.516 ± 0.14	1652496	65.11
2005	22.28 ± 0.78	0.542 ± 0.22	1546105	67.36
2006	22.34 ± 0.77	0.594 ± 0.11	1844239	68.02

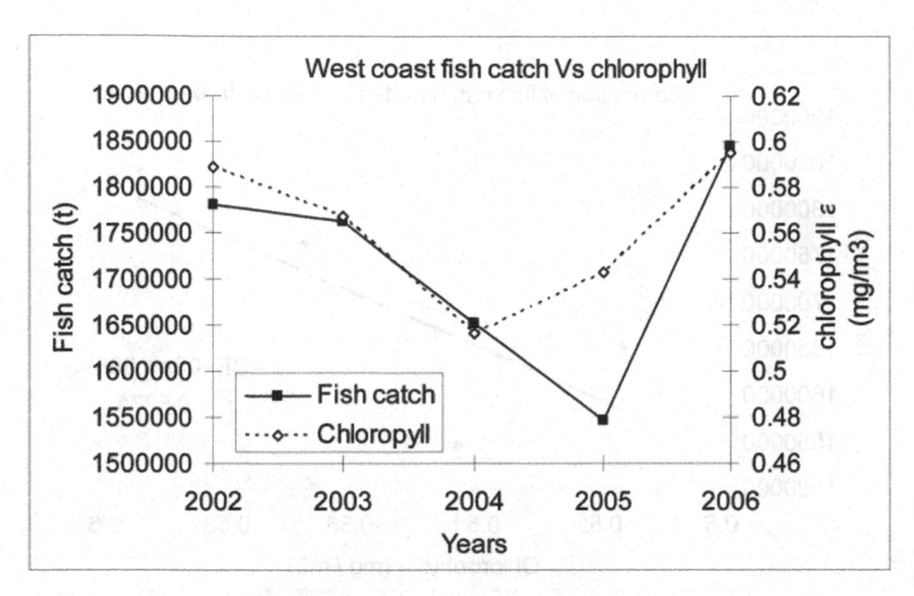

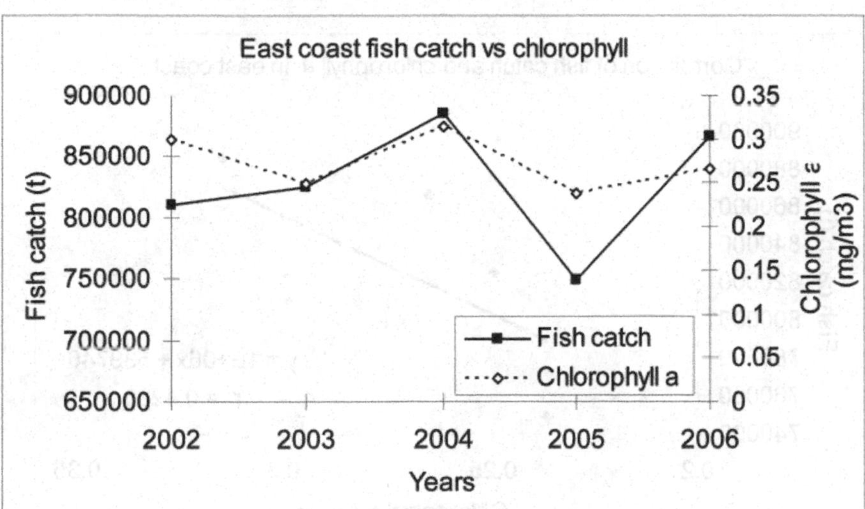

FIGURE 3.7 Mean fish catch and Chl *a* of east and west coast of India during 2002–2006.

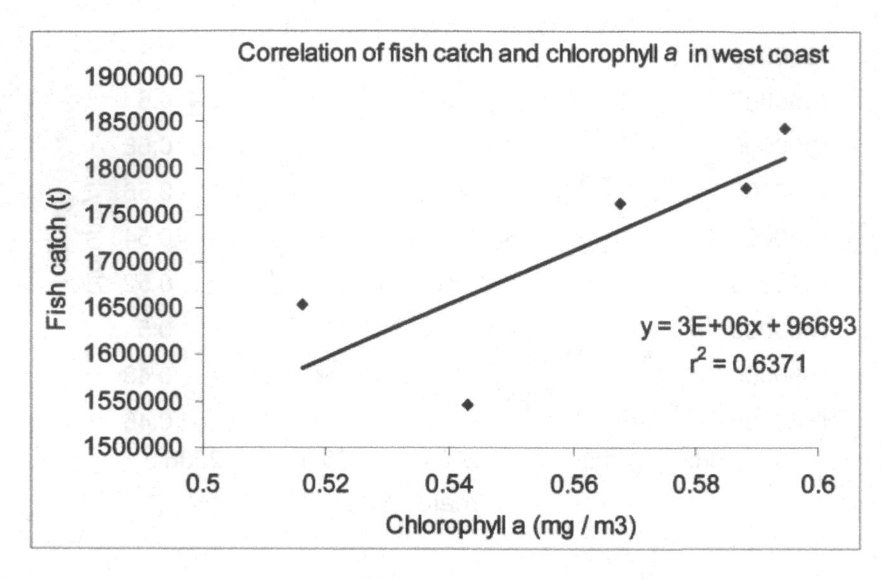

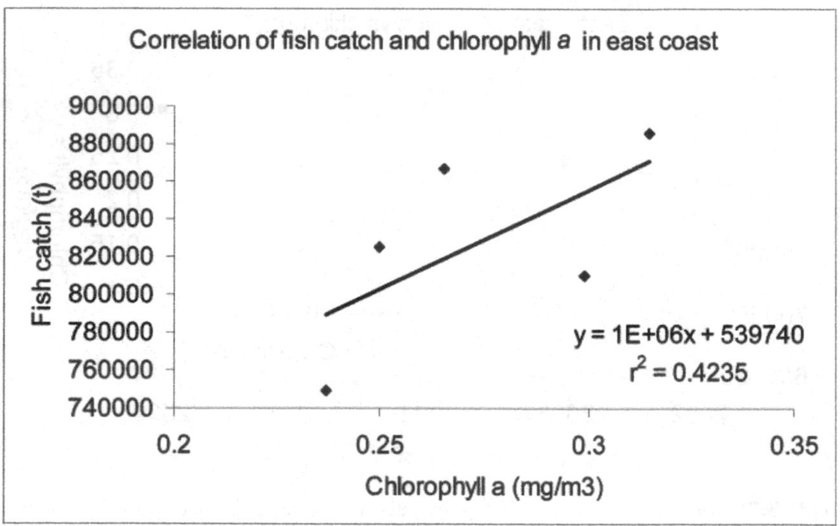

FIGURE 3.8 Correlation of fish catch and Chl *a* derived from MODIS.

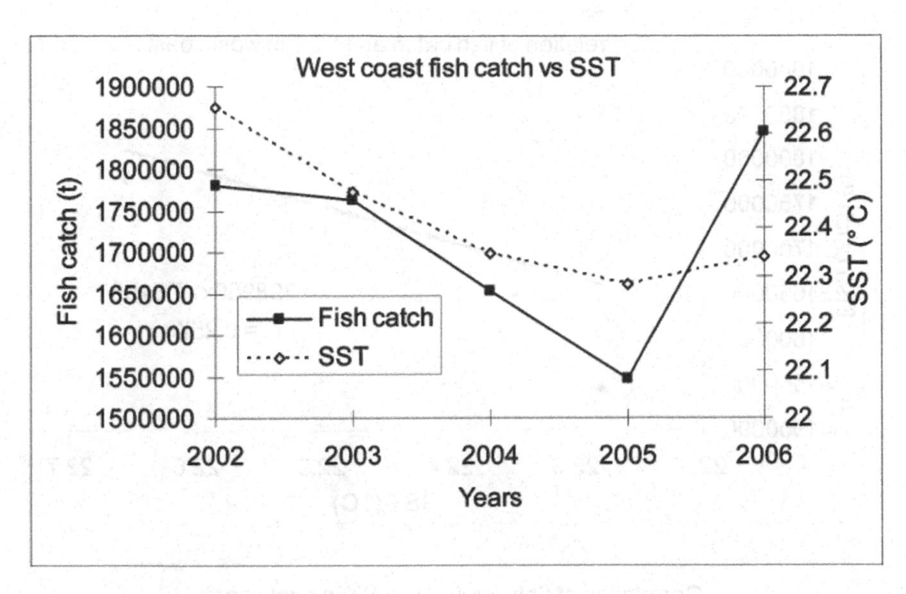

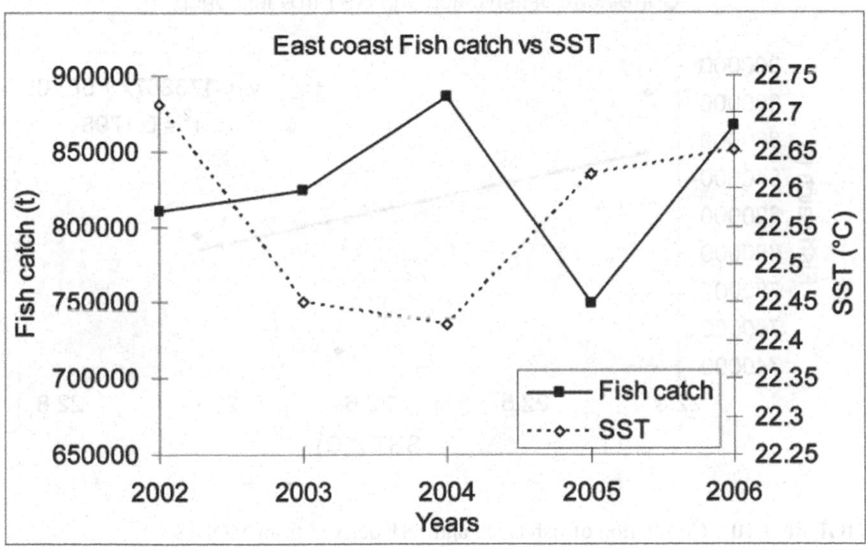

FIGURE 3.9 Mean fish catch and SST of east and west coast of India during 2002–2006.

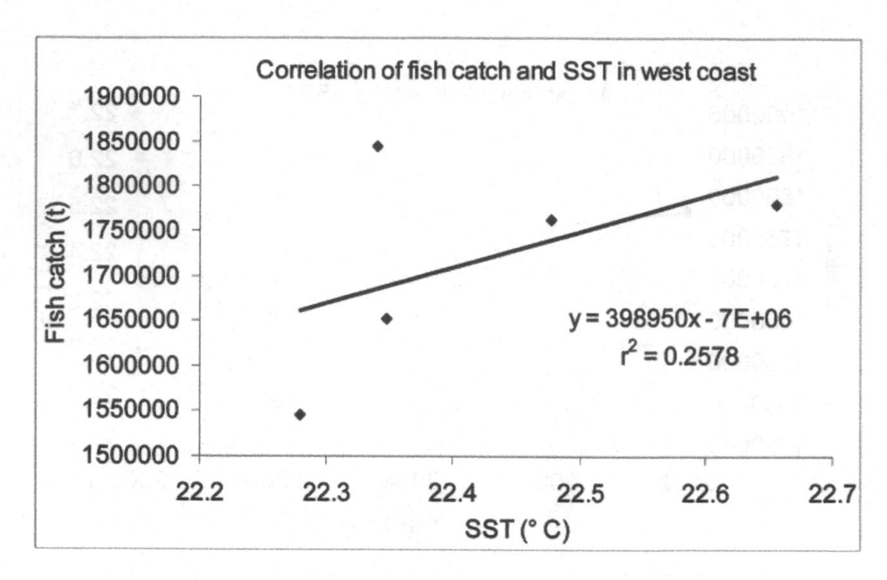

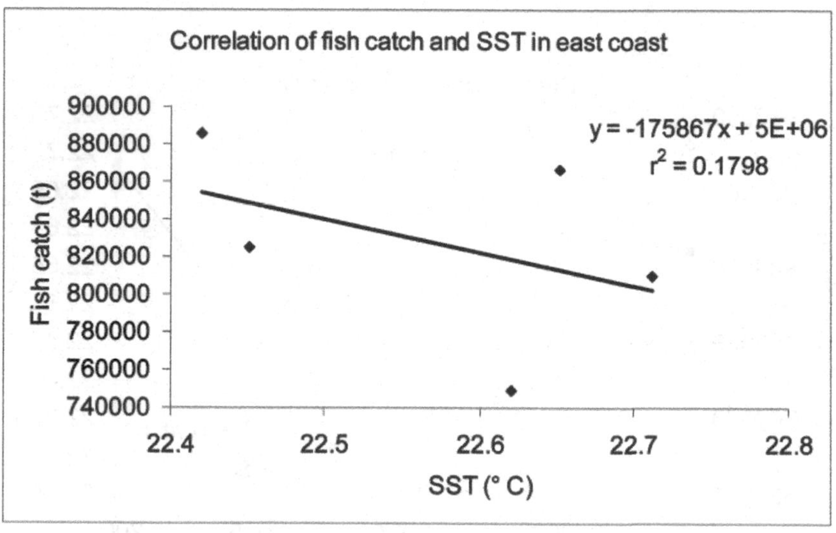

FIGURE 3.10 Correlation of fish catch and SST derived from MODIS.

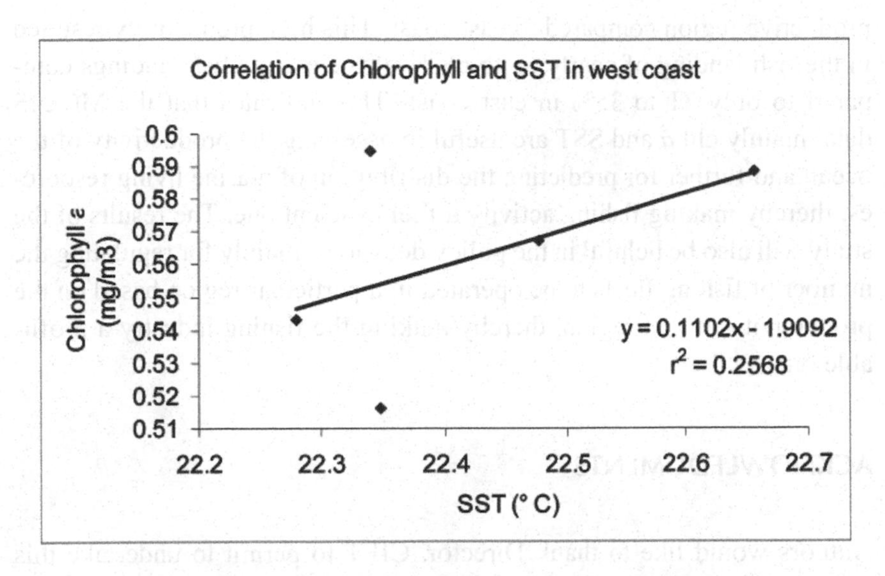

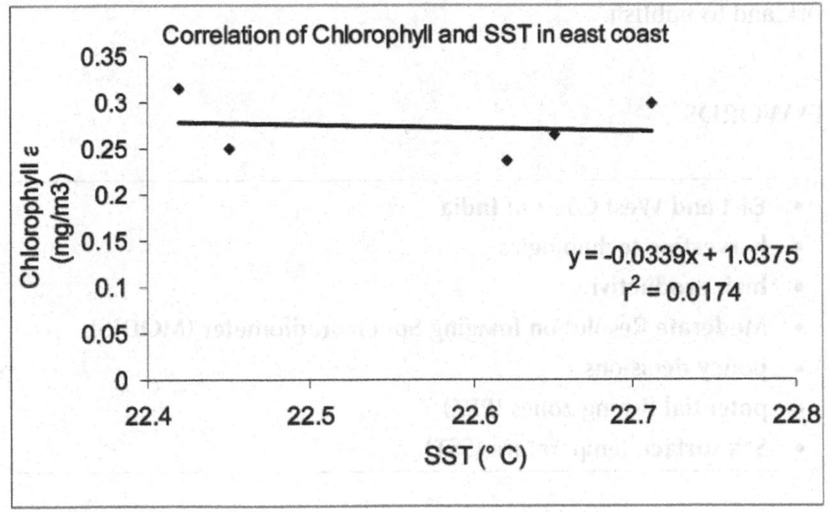

FIGURE 3.11 Correlation of SST and Chl *a* derived from MODIS.

3.4 CONCLUSION

The study indicated that the chl *a* concentration of west coast was twice higher than that of the east coast making the west coast of India a highly productive region compared to east coast. This high productivity resulted in the fish landing of nearly 65 to 68 % of the marine fish landings compared to only 31 to 35% in east coast. This indicates that the MODIS data, mainly chl *a* and SST are useful in assessing the productivity of the ocean and further for predicting the distribution of marine living resources, thereby making fishing activity a fuel efficient one. The results of the study will also be helpful in the policy decisions, mainly for regulating the number of fishing fleets to be operated in a particular region based on the productivity of that region, thereby making the fishing industry a profitable one.

ACKNOWLEDGMENTS

Authors would like to thank Director, CIFT to permit to undertake this work and to publish.

KEYWORDS

- **East and West Coast of India**
- **harvesting technologies**
- **high productivity**
- **Moderate Resolution Imaging Spectroradiometer (MODIS)**
- **policy decisions**
- **potential fishing zones (PFZ)**
- **Sea surface temperature (SST)**

REFERENCES

1. Froese, R. and Pauly, D., FishBase 2000: Concepts, Design and Data Sources. ICLARM, Los Banos, Philippines, 2000, 346 pp.
2. Watanabe, H. and Uchida, J., An estimation of direct and indirect energy input in catching fish for fish paste products. *Bull. Jpn. Soc. Sci. Fish.*, 1984, 50, 417–423

3. Rawitscher, M. A., Energy Cost of Nutrients in the American Diet. PhD Thesis, 1978, University of Connecticut, Storrs, Connecticutt.
4. Tyedmers, P., Salmon and Sustainability: The Biophysical Cost of Producing Salmon Through the Commercial Salmon Fishery and the Intensive Salmon Culture Industry. PhD Thesis, 2000, University of British Columbia, Vancouver, Canada.
5. Tyedmers, P., Energy consumed by North Atlantic fisheries. In Fisheries' Impacts on North Atlantic Ecosystems: Catch, Effort and National/Regional Datasets. Zeller, D., Watson, R. and Pauly, D. (eds.). Fisheries Centre, University of British Columbia, Vancouver, 2001, pp. 12–34.
6. Watanabe, H. and Okubo, M., Energy Input in Marine Fisheries of Japan. *Bull. Jpn. Soc. Sci. Fish.*, 1989, 53, 1525–1531.
7. Tyedmers, P., Fisheries and energy use. In Encyclopaedia of Energy. Cleveland, C. (ed.). Elsevier, San Diego, vol. 2. 2004, pp. 683–693.
8. Tyedmers, P. H. Watson, R. and Pauly, D., Fueling Global Fishing Fleets. *Ambio.*, 2005, 34 (8): 635–638.
9. FAO., Detailed annual species-specific catches based in large measure on data in FAO 2003 FishStat Plus statistical database, 2003. (http://www.fao.org/fi/statist/FISOFT/FISHPLUS.asp).
10. Sundby, S., Recruitment of Atlantic cod stocks in relation to temperature and advection of copepod populations. *Sarsia.*, 2000, 85: 277–298.
11. Kennish, M. J., Ecology of Estuaries, Volume II, 1986, Biological Aspects (Boca Raton: CRC Press).
12. Day, J. W. Jr., Hall, C.A.S., Kemp, W.M. and Ya' Nez-Arancibia, A., Estuarine Ecology, 1989, New York: John Wiley & Sons.
13. Boney, A. D., Phytoplankton, 1988, London: Edward Arnold.
14. Martin, S., An Introduction to Ocean Remote Sensing: Cambridge University Press, 2004, 426 p.
15. Han, L., Spectral reflectance with varying suspended sediment concentrations in clear and algal-laden waters. *Photogrammetric Engineering and Remote Sensing.*, 1997, 63, 701–705.
16. Feldman, G. C. Esaias, W. McClain, C. Elrod, J. Maynard, N. and Endres, D., Ocean Color: Availability of the global dataset, *EOS Trans. AGU.*, 1989, 70, 634–641.
17. McClain, C. R. Feldman, G. C. and, Esaias, W. E., Oceanic biological productivity. In Atlas of Satellite Observations Related to Global Change, R. Gurney, J. L. Foster, and C. L. Parkinson, Eds. Cambridge, U.K.: Cambridge Univ. Press, 1993, pp. 251–263.
18. Yoder, J. A. McClain, C. R. Feldman, G. C. and Esaias W. E., Annual cycles of phytoplankton chlorophyll concentrations in the global ocean: A satellite view. *Global Biogeochem. Cycles.*, 1993, 7, 181–193.
19. Esaias, W. E., Abbott, M. R., Barton, I., Brown, O. B., Campbell, J. W., Carder, K. L., Clark, D. K., Evans, R. H., Hoge, F. E., Gordon, H. R., Balch, W. M., Letelier, R., Minnett, P. J., An overview of MODIS capabilities for ocean science observations. *IEEE Trans. Geosci. Remote Sens.*, 1998, 36, 1250–1265.
20. GOI., Annual Report. Department of Animal Husbandry, Dairying & Fisheries, Ministry of Agriculture, Government of India, New Delhi,. 2007–08, p 115.
21. Subramanian, V., Sediment load of Indian Rivers, *Curr. Sci.*, 1993, 64, 928–930.

22. Bhargava, R. M. S., India's Exclusive Economic Zone (ed. Qasim, S. Z. and Roonal, G. S.), Omega Scientific, New Delhi, 1996, pp. 122–131.
23. Desai, B. N., Bhargava, R. M. S. and Sarupria, J. S., Estimates of fishery potentials of the EEZ of India. *Estuarine Coastal Shelf Sci.*, 1990, 30, 635–639.
24. Central Marine Fisheries Research Institute (CMFRI)., *Mar. Fish. Infor. Serv., Technical and Extension series.*, 1995, 136, 30.
25. Prasanna Kumar, S., Madhupratap, M., Dileep Kumar, M., Gauns, M., Muraleedharan, P. M., Sarma, V. V. S. S., De Souza, S. N., Physical control of primary productivity on seasonal scale in central and eastern Arabian Sea. *Proc. Ind. Ac.Sci (Earth and Planetary Science).*, 2000, 109, 433–442.
26. Prasanna Kumar, S., M. Madhupratap, M. Dileep Kumar, P. M. Muraleedharan, S. N. de Souza, M. Gauns and Sarma, V.V. S. S., High biological productivity in the central Arabian Sea during summer monsoon driven by Ekman pumping and lateral advection, *Curr. Sci.*, 2001, 81, 1633–1638.
27. Nakamoto, S., Prasanna Kumar, S., Oberhuber, J. M., Muneyama, K., and Frouin, R., Chlorophyll modulation of sea surface temperature in the Arabian Sea in a mixed-layer isopycnal general circulation model. *Geophy. Res. Lett.*, 2000, 27(6), 747–750.
28. Madhupratap, M., Prasanna Kumar, S., Bhattathiri, P. M. A., Kumar, M. D., Raghu Kumar, S., Nair, K. K. C. and Ramaiah, N., Mechanism of the biological response to winter cooling in the North-eastern Arabian Sea. *Nature.*, 1996, 384, 549–552.
29. Tang, D. L., Kawamura, H., and Luis, A. J., Short-term variability of Phytoplankton blooms associated with a cold eddy in the north-eastern Arabian Sea. *Remote Sens. Environ.*, 2002, 81, 82–89.
30. Gomes, H. R., J. I. Goes and Saino, T., Influence of physical processes and freshwater discharge on the seasonality of phytoplankton regime in the Bay of Bengal. *Cont. Shelf Res.*, 2000, 20, 313–330.
31. Prasanna Kumar, S., M. Muraleedharan, T. G. Prasad, M. Gauns, N. Ramaiah, S. N. De Souza, S. Sardesai and Madhupratap, M., Why is the Bay of Bengal less productive during summer monsoon compared to the Arabian Sea?. *Geophys. Res. Lett.*, 2002, 29(24), 2235.
32. Dey, S. and Singh, R. P., Comparison of chlorophyll distributions in the north-eastern Arabian Sea and southern Bay of Bengal using IRS-P4 Ocean Color Monitor data. *Remote Sen. Environ.*, 2003, 85, 424–428.
33. Vinayachandran, P. N. and Mathew, S., Phytoplankton bloom in the Bay of Bengal during the north-east monsoon and its intensification by cyclones. *Geophys. Res. Lett.*, 2003, 30(11), 1572.
34. Sathyendranath, S., Gouveia, A. D., Shetye, S. R., and Platt, T., Biological controls of surface temperature in the Arabian Sea. *Nature.*, 1991, 349, 54–56.
35. Sarma, V. V. Sadhuram, Y. Sravanthi, N. A. and Tripathy. S. C., Role of physical processes in the distribution of chlorophyll a in the north-west Bay of Bengal during pre and postmonsoon seasons. *Curr. Sci.*, 2006, 91(9), 1133–1134.

CHAPTER 4

MECHANISMS OF CATALYSIS WITH BINARY AND TRIPLE CATALYTIC SYSTEMS

L. I. MATIENKO, V. I. BINYUKOV, L. A. MOSOLOVA, E. M. MIL, and G. E. ZAIKOV

CONTENTS

4.1 Introduction 162
4.2 Experimental 163
4.3 Results and Discussion 164
4.4 Conclusion 184
Keywrods 186
References 186

4.1 INTRODUCTION

Mechanisms of catalysis with binary, and triple catalytic systems $\{Ni^{II}(acac)_2+L^2\}$ (L^2=MP,MP=N-metylpirrolidon-2), $\{Ni^{II}(acac)_2$+NaSt(or LiSt)+PhOH}, $\{Fe^{III}(acac)_3+L^2\}$ (L^2 = 18C6, CTAB) in the selective ethyl benzene oxidation by dioxygen into α-phenyl ethyl hydro peroxide are briefly discussed. The possibility of the formation of stable supramolecular nanostructures on the basis of nickel and iron heteroligand complexes $Ni^{II}(acac)_2$·NaSt(or LiSt)·PhOH, $Ni_2(OAc)_3(acac)L^2{\cdot}2H_2O$, $Fe_x(acac)_y18C6_m(H_2O)_n$, and $Fe_x(acac)_y(CTAB)_p(H_2O)_q$, due to intermolecular H-bonds is researched with the AFM method. The specific structure self organization of nickel and iron heteroligand complexes as models of Ni(Fe) ARD Dioxygenases may be used for understanding the actions of these enzymes.

Nanostructure science and supramolecular chemistry are fast evolving fields that are concerned with manipulation of materials that have important structural features of nanometer size (1 nm to 1 μm) [1, 2]. Nature has been exploiting no covalent interactions for the construction of various cell components. For instance, microtubules, ribosomes, mitochondria, and chromosomes use mostly hydrogen bonding in conjunction with covalently formed peptide bonds to form specific structures.

H-bonding can be a remarkably diverse driving force for the self-assembly and self-organization of materials. H-bonds are commonly used for the fabrication of supramolecular assemblies because they are directional and have a wide range of interactions energies that are tunable by adjusting the number of H-bonds, their relative orientation, and their position in the overall structure. H-bonds in the center of protein helices can be 20 kcal/mol due to cooperative dipolar interactions [3].

The porphyrin linkage through H-bonds is the binding type generally observed in nature. One of the simplest artificial self-assembling supramolecular porphyrin systems is the formation of a dimer based on carboxylic acid functionality (Fig. 4.1) [4].

The mechanism of catalysis often involves the formation of a supramolecular assembly during the reaction [4]. New approach have been offered to the research of mechanism of catalysis with bi- and triple-catalytic systems $\{Ni^{II}(acac)_2+L^2\}$ (L^2 = MP, MP=N-metylpirrolidon-2),

{$Ni^{II}(acac)_2$+NaSt(or LiSt)+PhOH}, {$Fe^{III}(acac)_3+L^2$} (L^2 = 18C6, CTAB) in the selective ethyl benzene oxidation by dioxygen into α-phenyl ethyl hydro peroxide. Stability of complexes $Ni^{II}(acac)_2$·NaSt(or LiSt)·PhOH, $Ni_2(OAc)_3(acac)L^2{\cdot}2H_2O$, $Fe_x(acac)_y18C6_m(H_2O)_n$, and $Fe_x(acac)_y(CTAB)_p(H_2O)_q$, which are the real catalytic particles in this process, – and also changes in mechanism of catalysis in the course of ethyl benzene oxidation, seems to be due to the supramolecular structures formation with assistance of intra and intermolecular H-bonds. The possibility of the formation of stable supramolecular nanostructures on the basis of nickel and iron heteroligand complexes due to intermolecular H-bonds we have researched with the AFM method.

4.2 EXPERIMENTAL

AFM SOLVER P47/SMENA/with Silicon Cantilevers NSG11S (NT MDT) with curvature radius 10 nm, tip height: 10–15 μm and cone angle £ 22° in taping mode on resonant frequency 150 KHz was used.

As substrate the polished Silicone surface special chemically modified was used. Modified Silicone surface was exploit for the self-assembly driven growth due to H-bonding of complexes $Ni^{II}(acac)_2$·NaSt(or LiSt)·PhOH, heteroligand complexes $Ni_2(AcO)_3(acac)MP{\cdot}2H_2O$, $Fe_x(acac)_y18C6_m(H_2O$, or $CHCl_3)_n$ with Silicone surface. The saturated solution of complexes was put on a surface, maintained some time, and then solvent was deleted from a surface by means of special method – spin-coating process.

In the course of scanning of investigated samples it has been found, that the structures are fixed on a surface strongly enough due to H-bonding. The self-assembly driven growth of the supramolecular structures on the basis of complexes, mentioned above, due to H-bonds and perhaps the other non-covalent interactions was observed on Silicone surface.

One can watch the structures on the basis of $Ni^{II}(acac)_2$·NaSt·PhOH with big height and volume. In check experiments it has been shown that for binary systems {$Ni^{II}(acac)_2$+NaSt}, and {$Ni^{II}(acac)_2$+PhOH}, the formation of the similar structures (exceeding on height of 2–10 nanometers) is not observed.

Ethyl benzene (RH) was oxidized with dioxygen at 120°C in glass bubbling-type reactor in the presence of three-component systems {$Ni^{II}(acac)_2$+L^2+PhOH} (L^2=NaSt, LiSt)) [4].

Analysis of oxidation products. α-Phenyl ethyl hydro peroxide (PEH) was analyzed by iodometry. By-products, including methylphenylcarbinol (MPC), acetophenone (AP), and phenol (PhOH) as well as the RH content in the oxidation process were examined by GLC [5, 6].

The order in which PEH, AP, and MPC formed was determined from the time dependence of product accumulation rate rations at $t \to 0$. The variation of these rations with time was evaluated by graphic differentiation [5, 6] (*see* Fig. 4.3).

Experimental data processing was done using special computer programs Mathcad and Graph2Digit.

4.3 RESULTS AND DISCUSSION

4.3.1 MECHANISM OF CATALYSIS WITH TRIPLE SYSTEMS {$NI(II)(ACAC)_2$+L^2+PHOH}

4.3.1.1 ROLE OF INTERMOLECULAR H-BONDING

The problem of selective oxidation of alkylarens to hydro peroxides is economically sound. Hydro peroxides are used as intermediates in the large-scale production of important monomers. For instance, propylene oxide and styrene are synthesized from α-phenyl ethyl hydro peroxide, and cumyl hydro peroxide is the precursor in the synthesis of phenol and acetone [7, 8]. The method of modifying the Ni^{II} and $Fe^{II, III}$ complexes used in the selective oxidation of alkylarens (ethyl benzene and cumene) with molecular oxygen to afford the corresponding hydro peroxides aimed at increasing their selectivity's has been first proposed by L.I. Matienko [5, 9]. This method consists of introducing additional mono- or multidentate modifying ligands into catalytic metal complexes. The mechanism of action of such modifying ligands was elucidated. New efficient catalysts of selective oxidation of ethyl benzene to α-phenyl ethyl hydro peroxide were developed [5, 9].

The phenomenon of a substantial increase in the selectivity (*S*) and conversion (*C*) of the ethyl benzene oxidation to the to α-phenyl ethyl hydro peroxide upon addition of PhOH together with alkali metal stearate MSt (M = Li, Na) as ligands to metal complexes $Ni^{II}(acac)_2$ was discovered in Refs. [5, 9] works.

The observed values of ***C*** [*C* >35% at $(S_{PEH})_{max}$ ~ 90%], $[ROOH]_{max}$ (1.6–1.8 mol $liter^{-1}$)] far exceeded those obtained with other ternary catalytic systems {$Ni^{II}(acac)_2$+ L^2 + PhOH} (L^2 is N-metylpyrrolidone-2 (MP), hexamethylphosphorotriamide (HMPA),) and the majority of active binary systems (Fig. 4.1).

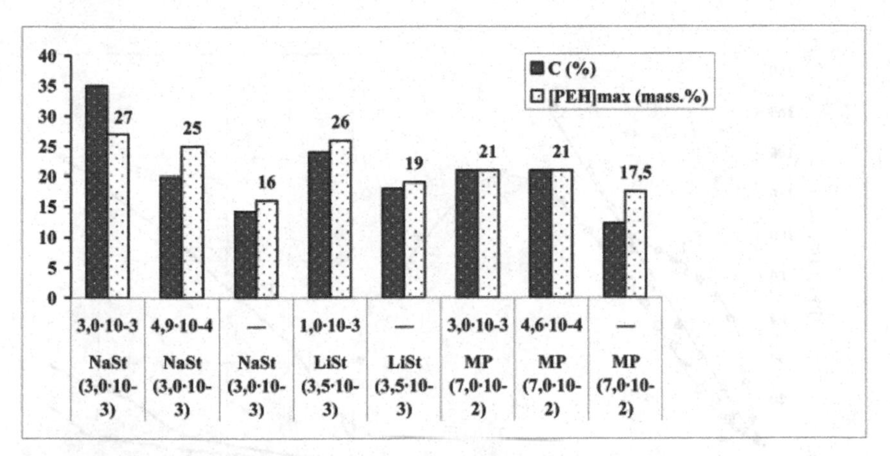

FIGURE 4.1 Values of conversion ***C*** (%) (I row), maximum values of hydro peroxide concentrations $[PEH]_{max}$ (mass.%) (II row) in reactions of ethyl benzene oxidation in the presence of triple catalytic systems {$Ni(II)(acac)_2$+L^2+PhOH} (L^2 = NaSt, LiSt, MP: [PhOH] (mol/l) – on an axis of abscises (the top number), [L^2] (mol/l) – on an axis of abscises (the bottom number)). [$Ni^{II}(acac)_2$] = 3.0×10^{-3} mol/l, 120°C.

There are characteristic features for triple systems including metalloligand–modifier L^2 = NaSt, LiSt (and N-metylpirrolidon-2) compared with the most active binary systems (Fig. 4.2). The advantage of these ternary systems is the long-term activity of the *in situ* formed complexes $Ni^{II}(acac)_2$•L^2•PhOH: unlike binary systems the acac ligand in nickel complex does not undergo transformations in the course of ethyl benzene oxidation. (The formation of triple complexes $Ni^{II}(acac)_2 \cdot L^2 \cdot PhOH$ at vary early stages of oxidation was established with kinetic methods [9]). So the reaction rate remains practically the same during the oxidation process. In the course of the oxidation the rates of products accumulation unchanged during the long period $t \leq 30–40$ hours (as one can see on Fig. 4.2 (a, b)).

At catalysis with $\{Ni^{II}(acac)_2 + MSt + PhOH\}$ the oxidation products PEH, AP, MPC are formed practically without the auto acceleration period (NaSt) or with maximum initial rate (LiSt) (unlike catalysis with triple system which does not contain metalloligand (L^2=MP)). Throughout the process of ethyl benzene oxidation the rates of formation of PEH (and AP, MPC also): w_P=const=max (P = PEH, AP, or MPC) (NaSt); w_P=const=max (P =AP, MPC), w_P=const ($<w_{max}$) (P=PEH) in the case of L^2 =LiSt (Fig. 4.2).

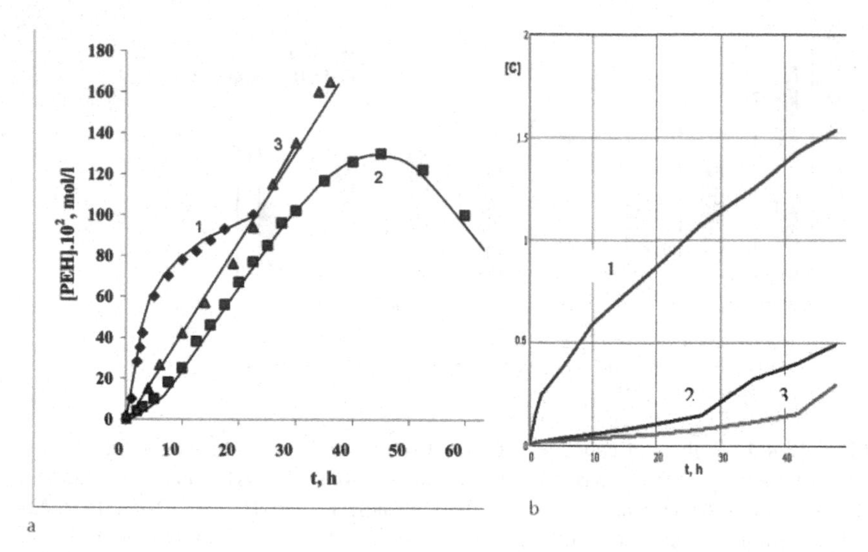

FIGURE 4.2 (a) Kinetics of accumulation of PEH in reactions of ethyl benzene oxidation, catalyzed by binary system $\{Ni^{II}(acac)_2+MP\}$ (1) and two triple systems $\{Ni^{II}(acac)_2+L^2+PhOH\}$ with L^2=MP (2) and L^2=NaSt (3). $[Ni^{II}(acac)_2] = 3 \cdot 10^{-3}$ mol/L, and $[MP] = 7 \cdot 10^{-2}$ mol/L, $[NaSt] = 3 \cdot 10^{-3}$ mol/L, $[PhOH] = 3 \cdot 10^{-3}$ mol/L, 120°C.

(b) Kinetics of accumulation of PEH (1), AP (2), MPC (3) in ethyl benzene oxidation, catalyzed by triple system $\{Ni^{II}(acac)_2+LiSt+PhOH\}$. $[Ni^{II}(acac)_2] = 3 \cdot 10^{-3}$ mol/L, [LiSt] $= 3 \cdot 10^{-3}$ mol/L, $[PhOH] = 3 \cdot 10^{-3}$ mol/L, 120°C (Data in Fig. 4.2b. are presented for the first time).

Often metals of constant valency compounds are used in combination with redox-active transition-metal complexes to promote a variety of reactions involving the transfer of electrons [10]. This effect is typified in metalloproteins such as the copper zinc super oxide dismutase, in which both metal ions have been proposed to be functionally active [10].

Earlier we have established, that the increase in the initial rate of the ethyl benzene oxidation with dioxygen, catalyzed with in the presence of additives of metalloligands MSt (M=Li, Na, K), is due to higher activity of formed complexes $Ni^{II}(acac)_2$·MSt in the micro stages of chain initiation and/ or decomposition of PEH with free radical formation [5, 9].

At that the participation of catalyst NiII(acac)2•MSt in micro steps of chain propagation and, probably, in chain termination must also be taken into account [5, 9]. The results found in Ref. [10] illustrate the possibility that redox-inactive metal ions can be used to facilitate the activation of dioxygen, which are consistent with our data.

At catalysis with triple complexes Ni(acac)2•L2•PhOH (L2 = NaSt, LiSt) the parallel formation of a-phenyl ethyl hydro peroxide, acetophenone and MPC was observed (wAP(MPC)/wPEH ¹ 0 at t ® 0, wAP/wMPC ¹ 0 at t ® 0) throughout the reaction of ethyl benzene oxidation) (see, for example, Fig. 4.3a, b).

A more considerable increase in the selectivity ($\boldsymbol{S}_{PEH}$) at the catalysis by $Ni^{II}(acac)_2$·L^2·PhOH (L^2 = NaSt, LiSt) complexes compared with non-catalytic oxidation and binary systems {$Ni^{II}(acac)_2$+MSt} was associated with the change in the route of acetophenone and methylphenylcarbinol formation: AP and MPC form in parallel with PEH rather than as a result of PEH decomposition, – and with the inhibition of the PEH heterolytic decomposition with PhOH formation.

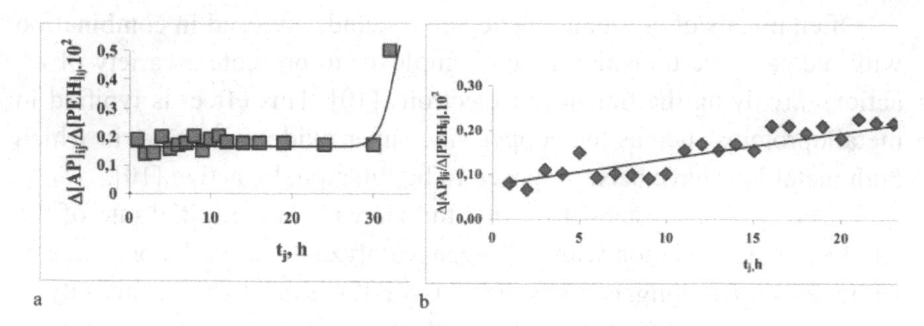

FIGURE 4.3 (a) Dependence Δ $[AP]_{ij}$/ Δ $[PEH]_{ij}\cdot 10^2$ on time t_j in the course of ethyl benzene oxidation, catalyzed with complexes $Ni^{II}(acac)_2\cdot NaSt\cdot PhOH$ (1:1:1), 120°C. (b) Dependence Δ $[AP]_{ij}$/ Δ $[PEH]_{ij}\cdot 10^2$ on time t_j in the course of ethyl benzene oxidation, catalyzed with complexes $Ni^{II}(acac)_2\cdot LiSt\cdot PhOH$ (1:1:1), 120°C.

Thus we had shown that the triple complexes NiII(acac)2·NaSt(LiSt)·PhOH unlike binary complexes NiII(acac)2·NaSt(LiSt) obviously were inactive in the reaction of hydro peroxide decomposition. However the ability of redox-inactive metal ions to facilitate the activation of dioxygen (and free radical formation in chain initiation), continued in the case of the catalysis with the triple complexes NiII(acac)2·NaSt(LiSt)·PhOH.

In these systems dioxygen activation may be promoted through the formation of intramolecular H-bonds [9]. The role of intramolecular H-bonds is established by us in mechanism of formation of triple catalytic complexes {NiII(acac)2·L2·PhOH} (L2 = N-methylpirrolidon-2) in the process of ethyl benzene oxidation with molecular oxygen [9]. As can be seen from Fig. 4.2, complexes NiII(acac)2·LiSt·PhOH, probably, have a higher activity in relation to molecular oxygen compared to complexes NiII(acac)2·NaSt(MP)·PhOH.

The high efficiency of three-component systems {$Ni^{II}(acac)_2$+MSt+PhOH} (M = Na, Li) in the reaction of selective oxidation of ethyl benzene to α-phenyl ethyl hydro peroxide was associated with the formation of extremely stable heterobimetallic heteroligand complexes $Ni(acac)_2\cdot MSt\cdot PhOH$. The stability of complexes $Ni^{II}(acac)_2\cdot MSt\cdot PhOH$ can be associated with the formation of intermolecular H-bonds [11–13]. In this paper we have applied AFM technique for research the possibility of supramolecular structures formation in the process of ethyl benzene oxidation to the α-phenyl ethyl hydro peroxide at catalysis with three-component systems:

$$\{Ni^{II}(acac)_2+NaSt(or\ LiSt)+PhOH\}\rightarrow Ni^{II}(acac)_2\cdot NaSt(or\ LiSt)\cdot PhOH\rightarrow\{Ni^{II}(acac)_2\cdot NaSt(or\ LiSt)\cdot PhOH\}_n$$

4.3.1.2 ROLE OF H-BONDING STABILIZATION OF TRIPLE CATALYTIC COMPLEXES $NI^{II}(ACAC)_2\cdot L^2\cdot PHOH$

The association of triple complexes $Ni^{II}(acac)_2\cdot NaSt$(or LiSt)·PhOH to supramolecular structures due to H-bonding should be followed from analysis of AFM data, which we received in our works. Results are presented on the Figs. 4.4–4.9 and Table 4.1. Data on the structures on the basis of $Ni^{II}(acac)_2\cdot LiSt\cdot PhOH$ complexes are presented for the first time.

Figures 4.4 and 4.5 demonstrated three-dimensional and two-dimensional AFM image (30×30 and 10×10(μm)) of the structures on the basis of triple complexes $Ni^{II}(acac)_2\cdot NaSt\cdot PhOH$ [14] formed at drawing of a uterine solution on a surface of modified silicone.

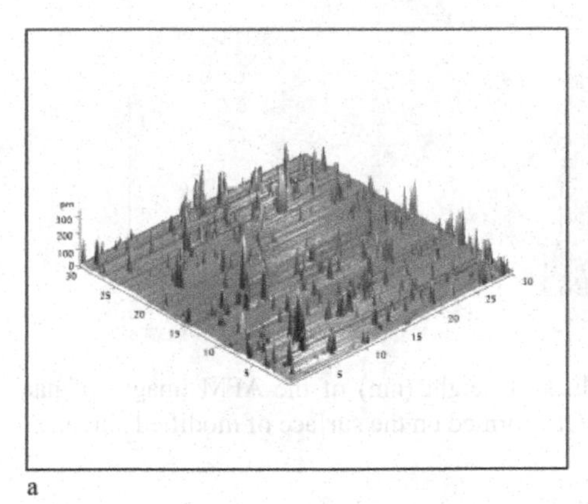

a

b

FIGURE 4.4 The AFM three-dimensional image (30×30 (a) and 10×10 (b) (μm)) of the structures formed on a surface of modified silicone on the basis of triple complexes $Ni^{II}(acac)_2\cdot NaSt\cdot PhOH$.

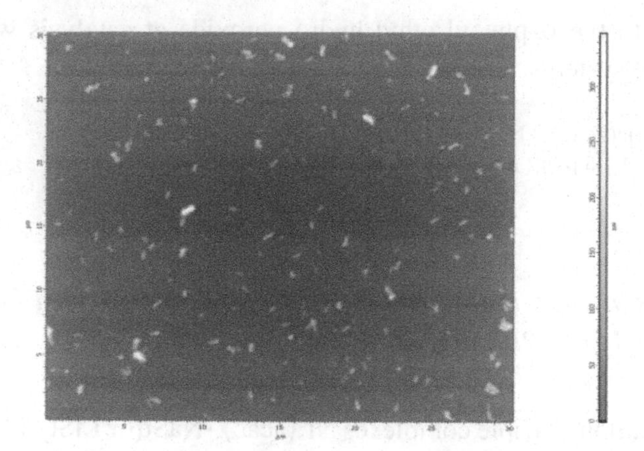

FIGURE 4.5 The AFM two-dimensional image (30×30 (μm)) of nanoparticles on the basis $Ni^{II}(acac)_2$·NaSt·PhOH formed on the surface of modified silicone.

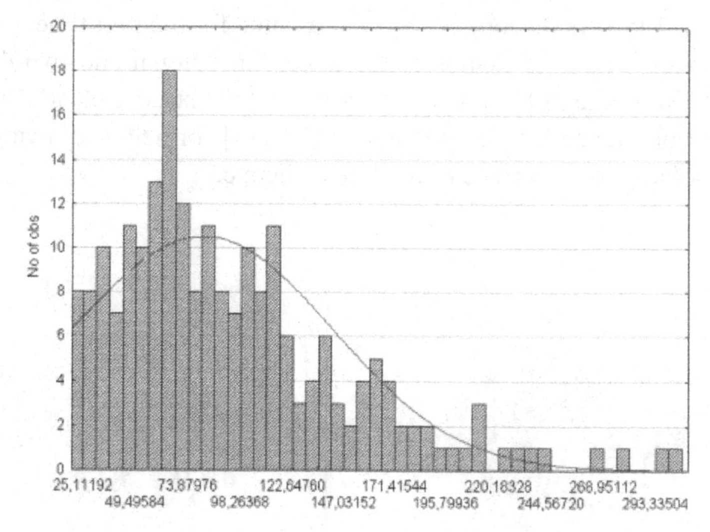

FIGURE 4.6 Histogram of mean values of height (nm) of the AFM images of nano structures based on $Ni^{II}(acac)_2$·NaSt·PhOH, formed on the surface of modified silicone.

In the Fig. 4.6, the histogram of mean height of nanoparticles on basis of $Ni^{II}(acac)_2$·NaSt·PhOH is presented. As can see, structures are various on heights from the 25 nm to ~250–300 nm for maximal values. The distribution histogram shows that the greatest number of particles – is particles of the mean size of 50–100 nm on height.

Table 4.1 shows the mean values of area, volume, height, width, length of nanoscale structures on basis of triple complexes $Ni^{II}(acac)_2$·NaSt·PhOH formed on the surface of modified silicone.

TABLE 4.1 The mean values of area, volume, height, length, width of the AFM image of nanoparticles on the basis of $Ni^{II}(acac)_2$·NaSt·PhOH formed on the surface of modified silicone.

Variable	Mean values	Confidence −95.000%	Confidence +95.000%
Area (μm²)	0.13211	0.11489	0.14933
Volume (μm³)	14.11354	11.60499	16.62210
Z (Height) (nm)	80.56714	73.23940	87.89489
Length (μm)	0.58154	0.53758	0.62549
Width (μm)	0.19047	0.17987	0.20107

We have revealed an interesting fact that the length of the formed nanoparticles in the XY plane exceeds the width of the nanoparticles about three times (Table 4.1).

On the next pictures (Figs. 4.7–4.9) the nanoparticles image on the basis of $Ni^{II}(acac)_2$·LiSt·PhOH and $Ni^{II}(acac)_2$·LiSt complexes is demonstrated.

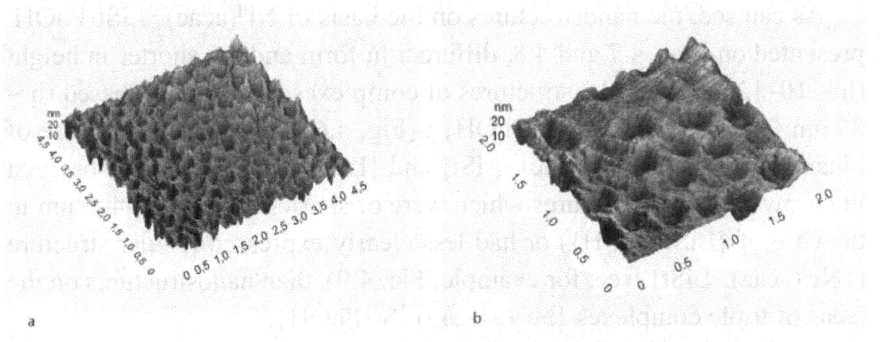

FIGURE 4.7 The AFM three-dimensional image (4.5×4.5 (a) and 2×2 (b) (μm)) of the structures formed on a surface of modified silicone on the basis of triple complexes $Ni^{II}(acac)_2$·LiSt·PhOH.

As one can see, nano structures on the basis of triple complexes $Ni^{II}(acac)_2$·LiSt·PhOH have the interesting cell form with cell height ~10 nm and cell width ~0.5 μm.

On the Fig. 4.8, three- and two-dimensional image (5×5 and 2×2 (μm)) and profile of one of the nanostructures on the basis of triple complexes $Ni^{II}(acac)_2$·LiSt·PhOH with more simple form (cell height ~7–12 nm and cell width ~60 nm), which we observed also on a surface of modified silicone, are presented.

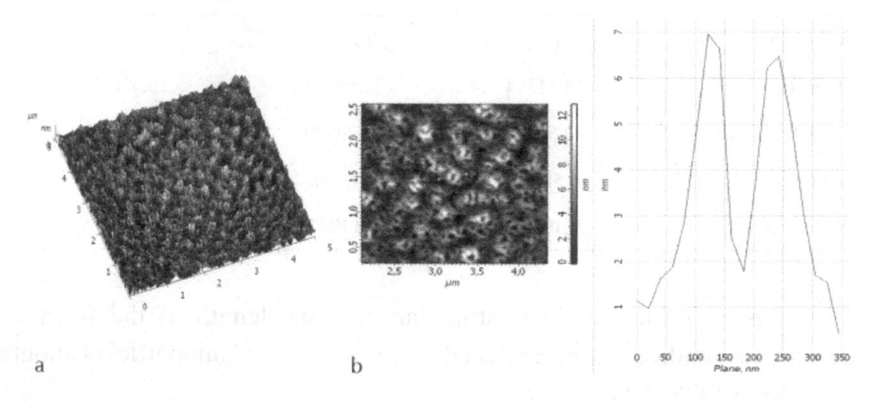

FIGURE 4.8 The AFM three- and two-dimensional image (5×5 (a) and 2.5×2.5 (b) (μm)) of the structures with more simple form received on a surface of modified silicone on the basis of triple complexes $Ni^{II}(acac)_2$·LiSt·PhOH and profile of one of these structures (c).

As can see, the nanostructures on the basis of $Ni^{II}(acac)_2$·LiSt·PhOH, presented on Figs. 4.7 and 4.8, different in form and are shorter in height (h ~ 10–12 nm) than the structures of complexes with sodium-based (h ~ 80 nm for $\{Ni^{II}(acac)_2\cdot NaSt\cdot PhOH\}_n$. (Fig. 4.6, Table 4.1). In the case of binary complexes, $\{Ni^{II}(acac)_2\cdot LiSt\}$ and {LiSt·PhOH}, we also observed the growth of nanostructures which were or shorter in height (~4–6 nm in the case of {LiSt·PhOH}) or had less clearly expressed regular structure ($\{Ni^{II}(acac)_2\cdot LiSt\}$ (see, for example, Fig. 4.9), than nanostructures on the basis of triple complexes $\{Ni^{II}(acac)_2\cdot LiSt\cdot PhOH\}$.

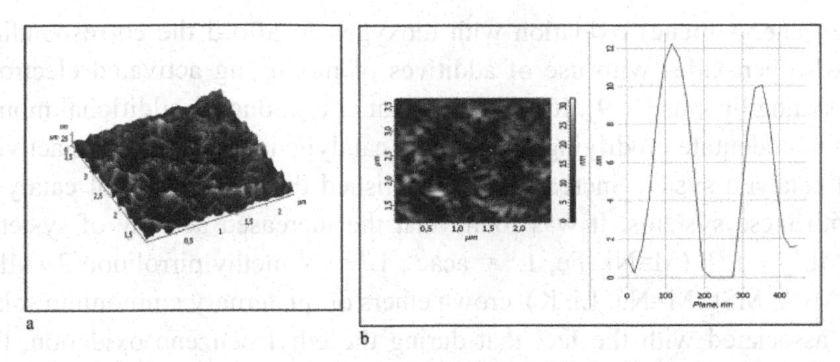

FIGURE 4.9 The AFM three-dimensional image (2.5×2.5 (μm)) (a), two-dimensional image (2.5×2.5 (μm)) and profile (b) of the nanostructures formed on a surface of modified silicone on the basis of $Ni^{II}(acac)_2 \cdot LiSt$ complexes.

So, we can conclude that the high degrees of conversion of ethyl benzene to PEH and the yields of α-phenyl ethyl hydro peroxide in the case of catalysis of three-component systems $\{Ni^{II}(acac)_2 + MSt + PhOH\}$ (M = Na, Li) in the reaction of selective oxidation of ethyl benzene to α-phenyl ethyl hydro peroxide, may be associated, as one of the reasons, with the formation of heterobimetallic heteroligand complexes $Ni^{II}(acac)_2 \cdot MSt \cdot PhOH$, which are self organized during ethyl benzene oxidation to extremely stable supramolecular structures $\{Ni^{II}(acac)_2 \cdot NaSt(\text{or } LiSt) \cdot PhOH\}_n$ at expense of intermolecular (phenol–carboxylate) H-bonds and, possible, the other noncovalent interactions [13]. The higher efficiency of heterobinuclear heteroligand complexes $Ni^{II}(acac)_2 \cdot NaSt \cdot PhOH$, including metalloligand NaSt, as selective catalysts compared with $Ni^{II}(acac)_2 \cdot LiSt \cdot PhOH$, seems to be due to formation of more stable supramolecular structures $\{Ni^{II}(acac)_2 \cdot NaSt \cdot PhOH\}_n$. The data in Figs. 4.4–4.9 show in favor of that assumption.

4.3.1.3 ROLE OF H-BONDING IN STABILIZATION OF TRANSITION CATALYTIC ACTIVE COMPLEXES $NI_X(ACAC)_Y(ACO)_Z(L^2)_N(H_2O)_M$ ("A"), $FE_X(ACAC)_Y 18C6_M(H_2O)_{N'}$, AND $FE_X(ACAC)_Y(CTAB)_P(H_2O)_Q$

In this work, we first proposed method for enhancing the catalytic activity of transition metal complexes in the processes of alkylarens (ethyl benzene, cumene) oxidation with dioxygen to afford the corresponding hydro peroxides with use of additives of modifying activated electron-donating ligands [5, 9]. It was found that at introducing additional mono- or multidentate modifying ligands into catalytic metal complexes activity of catalytic system increase. We established the mechanism of catalysis with these systems. It was found that the increased activity of systems $\{ML^1_n + L^2\}$ (M=Ni, Fe, L^1 = acac$^-$, L^2 = N-methylpirrolidon-2 (MP), HMPA, MSt (M=Na, Li, K), crown ethers or quaternary ammonium salts) is associated with the fact that during the ethyl benzene oxidation, the primary $(M^{II}L^1_2)_x(L^2)_y$ complexes (I macro stage) and the real catalytic active heteroligand $M^{II}_xL^1_y(L^1_{ox})_z(L^2)_n(H_2O)_m$ complexes (II macro stage) are formed to be involved in the oxidation process [5, 9]. We established the mechanism of formation of active heteroligand complexes $M^{II}_xL^1_y(L^1_{ox})_z(L^2)_n(H_2O)_m$ (M^{II}=Ni^{II}, Fe^{II}).

We have found that the electron-donating ligand L^2, axially coordinated to $M^{II}L^1_2$ (M=Ni, Fe, L^1 =acac$^-$), controls the formation of primary active complexes $M^{II}L^1_2{\cdot}L^2$ and the subsequent reactions in the outer coordination sphere of these complexes. The coordination of an electron-donating extraligand L^2 with an $M^{II}L^1_2$ complex, favorable for stabilization of the transient zwitter-ion L^2 $[L^1\,M(L^1)^+O_2^-]$, enhances the probability of regioselective O_2 addition to the methine C−H bond of an acetylacetonate ligand, activated by its coordination with metal ions. The outer-sphere reaction of O_2 incorporation into the chelate ring depends on the nature of the metal. Transformation routes of a ligand (acac)$^-$ for Ni and Fe are various, but lead to formation of similar heteroligand complexes $M^{II}_xL^1_y(L^1_{ox})_z(L^2)_n(H_2O)_m$ (L^1_{ox}=CH_3COO^-, M=Ni, Fe) [5, 9]. Thus for nickel complexes, the reaction of acac-ligand oxygenation follows a mechanism analogous to those of Ni^{II}-containing Acireductone Dioxygenase (ARD) [15, 17] or Cu- and Fe-containing Quercetin 2,3-Dioxygenases [18]. Namely, incorporation of O_2 into the chelate acac-ring was accompanied by the proton transfer and the redistribution of bonds in the transition complex leading to the scission of the cyclic system to form a chelate ligand OAc^-, acetaldehyde and CO (in the Criegee rearrangement).

In the effect of Fe^{II}-acetylacetonate complexes, we have found [5, 9] the analogy with the action of Fe^{II}-Acireductone Dioxygenase (ARD') [15, 17] and with the action of Fe^{II} containing Acetyl acetone Dioxygenase (Dke1)

[16]. For iron complexes oxygen adds to C–C bond (rather than inserts into the C=C bond as in the case of catalysis with nickel(II) complexes) to afford intermediate, i.e., a Fe complex with a chelate ligand containing 1,2-dioxetane fragment. The process is completed with the formation of the $(OAc)^-$ chelate ligand and methylglyoxal as the second decomposition product of a modified acac-ring (as it has been shown in [16]).

The methionine salvage pathway (MSP) (Scheme 1) plays a critical role in regulating a number of important metabolites in prokaryotes and eukaryotes. Acireductone dioxygenases (ARDs) Ni(Fe)-ARD are enzymes involved in the methionine recycle pathway, which regulates aspects of the cell cycle. The relatively subtle differences between the two metalloproteins complexes are amplified by the surrounding protein structure, giving two enzymes of different structures and activities from a single polypeptide (Scheme 1) [17]. Both enzymes $Ni^{II}(Fe^{II})$-ARD are members of the structural super family, known as cupins, which also include Fe-Acetyl acetone dioxygenase (Dke1) and Cysteine Dioxygenase. Structural and functional differences between the two ARDs enzymes are determined by thc type of metal ion bound in the active site of the enzyme.

One of the reasons for the different activity of $Ni^{II}(Fe^{II})$-ARD in the functioning of enzymes in relation to the common substrates (Acireductone (1,2-Dihydroxy-3-keto-5-methylthiopentene-2) and O_2) can be the association of catalyst in various macrostructure due to intermolecular H-bonds.

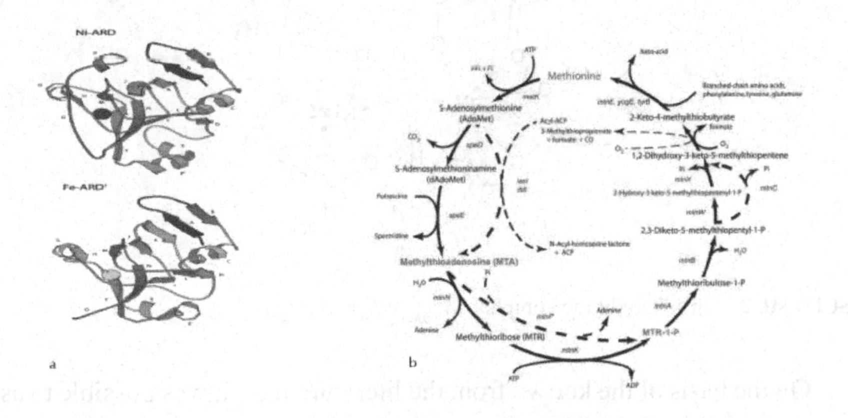

SCHEME 1 Acireductone dioxygenases Ni-ARD and Fe-ARD' (a) [17] are involved in the methionine recycle pathway (b).

4.3.1.4 ROLE OF H-BONDING IN STABILIZATION OF CATALYTIC COMPLEXES $NI_X(ACAC)_Y(OAC)_Z(L^2)_N(H_2O)_M$

It is known that heteroligand complexes are more active in relation to reactions with electrophiles in comparison with homoligand complexes [5, 9]. Thus the stability of heteroligand complexes $Ni_x(acac)_y(OAc)_z(L^2)_n(H_2O)_m$, the intermediate products of oxygenation of primary complexes $(M^{II}L^1{}_2)_x(L^2)_y$, with respect to conversion into inactive form, the end product of the complexes $(M^{II}L^1{}_2)_x(L^2)_y$ oxygenation: $Ni(OAc)_2$, – can be associated with the formation of stable macrostructures due to intermolecular H-bonds.

On possibility of supramolecular structures formation on the basis of heteroligand nickel complexes $Ni_x(acac)_y(OAc)_z(L^2)_n(H_2O)_m$ point data of AFM – spectroscopy, received in our works.

The complex, formed in the course of ethyl benzene oxidation, catalyzed with system $\{Ni^{II}(acac)_2 + MP\}$, has been synthesized by us and its structure has been defined with mass spectrometry, electron and IR spectroscopy and element analysis [5, 9]. The certain structure of a complex $Ni_{2(OAc)3(acac)}$·MP·$2H_2O$ corresponds to structure that is predicted on the basis of the kinetic data [5, 9]. Prospective structure of the complex $Ni_{2(OAc)3(acac)}$·MP·$2H_2O$ is presented with Scheme 2.

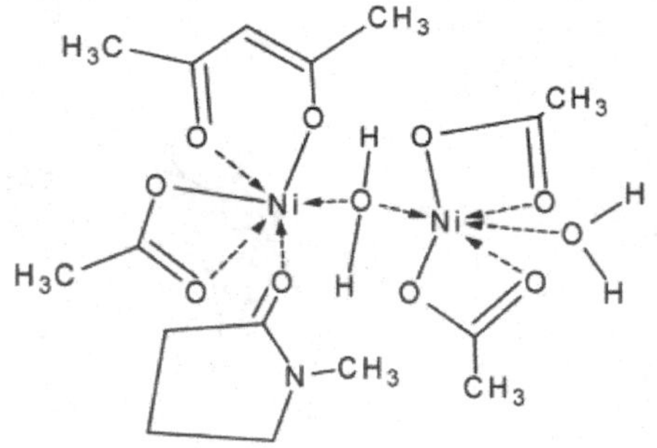

SCHEME 2 Structure of the complex $Ni_{2(OAc)3(acac)}$·MP·$2H_2O$

On the basis of the known from the literature facts it was possible to assume that heteroligand complexes $Ni_2(OAc)_3(acac)MP·2H_2O$ are capable to form macro structures with the assistance of intermolecular H-bonds (H_2O – MP, H_2O – (OAc^-)(or $(acac^-)$)) [12, 13].

The association of $Ni_{2(AcO)3(acac)}\cdot MP\cdot 2H_2O$ to supramolecular structures as result of H-bonding is demonstrated on the next Figs. 4.10–4.12.

On Figs. 4.10–4.12, three-dimensional and two-dimensional AFM image of the structures, formed at putting a uterine solution on a hydrophobic surface of modified silicone, is presented [19]. It is visible that the majority of the generated structures have rather similar form of three almost merged spheres.

As it is possible to see in Figs. 4.10a and 4.12, except particles with the form reminding three almost merged spheres (Fig. 4.10b), there are also structures of more than a simple form (with the height approximately equal 3–4 nm). A profile of one of the particles with size 3–4 nm on height is presented on Fig. 4.3.

The distribution histogram (Fig. 4.11) shows that the greatest number of particles – is particles with the size 3–4 nm on height.

From the data in Figs. 4.10–4.12, the following is visible. It is important to notice that for all structures the sizes in plane XY do not depend on height on *Z*. They make about 200 nm along a shaft, which are passing through two big spheres, and about 150 nm along a shaft crossing the big and smaller spheres (particles with the form reminding three almost merged spheres). But all structures are various on heights from the minimal 3–4 nm to ~ 20–25 nm for maximal values. In distribution on height there is a small quantity of particles with maximum height 20–25 nm and considerably smaller quantity with height to 35 nm (Fig. 4.10b).

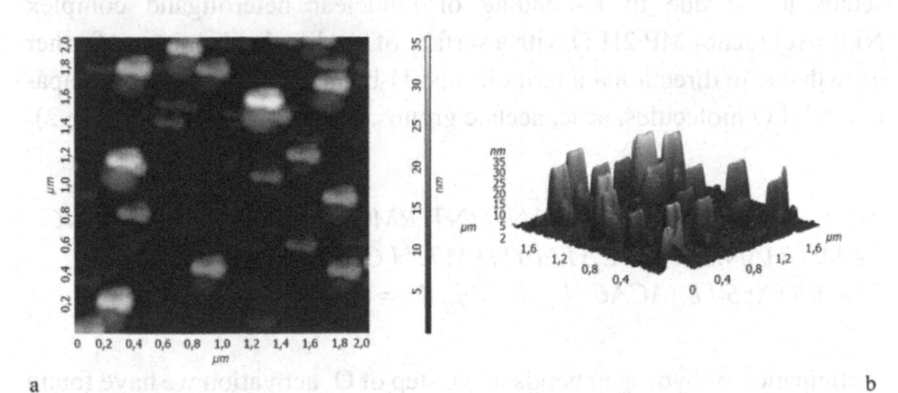

FIGURE 4.10 The AFM two- (a) and three-dimensional (b) image of nanoparticles on the basis $Ni_2(AcO)_3(acac)\cdot L^2\cdot 2H_2O$ formed on the hydrophobic surface of modified silicone.

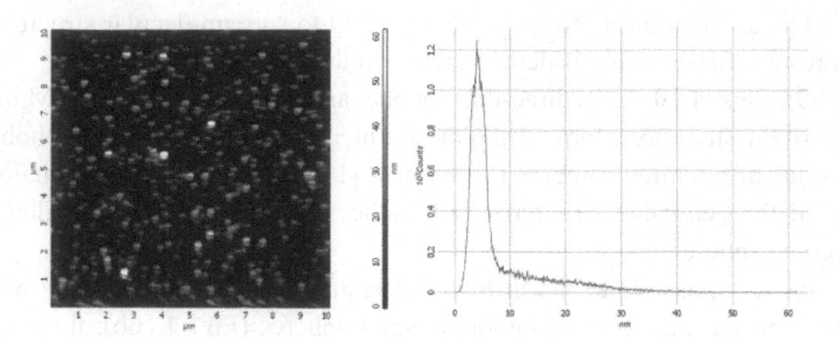

FIGURE 4.11 AFM image of structures on the hydrophobic modified silicone surface 10×10 μm (at the left). The distribution histogram on height of nanoparticles (to the right).

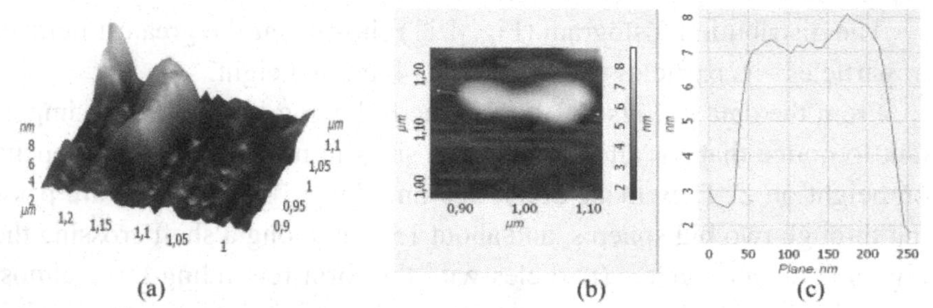

FIGURE 4.12 The AFM three- (a) and two-dimensional (b) image and profile of the structure (c) with minimum height along the greatest size in plane XY.

Thus, here we shows what the self-assembly driven growth seems to be due to H-bonding of binuclear heteroligand complex $Ni_2(OAc)_3(acac)\cdot MP\cdot 2H_2O$ with a surface of modified silicone, and further growth due to directional intermolecular H-bonds, apparently at participation of H_2O molecules, acac, acetate groups, MP [12, 13] (see Scheme 2).

4.3.1.5 ROLE OF INTRA AND INTERMOLECULAR H-BONDING IN MECHANISM OF CATALYSIS WITH CATALYTIC ACTIVE COMPLEXES $FE_X(ACAC)_YL^2{}_M(H_2O)_N$ (L^2 = 18C6, CTAB)

Participation of hydrogen bonds in the step of O_2 activation we have found in the oxidation of ethyl benzene with O_2 in the presence of catalytic systems $\{Fe^{III}(acac)_2 + L^2\}$ (L^2 = 18C6, CTAB, DMF). In the ethyl benzene oxidation of ion Fe^{III} rapidly transformed to Fe^{II} by the following reaction:

$$Fe^{III}(acac)_3\ ((Fe^{III}\ (acac)_2)_m{\cdot}(L^2)_n) + RH \circledR\ Fe^{II}(acac)_2\ ((Fe^{II}(acac)_2)_x{\cdot}(L^2)_y) \ldots \text{Hacac} + R^{\cdot}$$

Then the complex of Fe^{II} is involved in the chain initiation reaction (activation of O_2) and the reactions leading to the conversion of primary complexes $Fe^{II}(acac)_2)_x{\cdot}(L^2)_y$ in the active species $Fe^{II}_x L^1_y (L^1_{ox})_z (L^2)_n (H_2O)_m$ (through step of O_2 activation [16]). It was also found that complexes with HMPA, which do not form hydrogen bonds are not transforming into the active species $Fe^{II}_x L^1_y (L^1_{ox})_z (L^2)_n (H_2O)_m$ [5, 9]. The role of H-bonds in the mechanism of catalysis with chemical and biological systems follows from the AFM data presented below.

The surrounding protein structure, give two enzymes $Ni^{II}(Fe^{II})$ARD of different structures and activities. Association of the catalyst in macro-structures with the assistance of the intermolecular H-bonds may be one of reasons of reducing Ni^{II}ARD activity in mechanisms of $Ni^{II}(Fe^{II})$ARD operation [17]. On the other hand the Fe^{II}ARD operation seems to comprise the step of oxygen activation ($Fe^{II}+O_2 \rightarrow Fe^{III}{-}O_2^{-}{\cdot}$) (by analogy with Dke1 action [16]). Specific structural organization of iron complexes may facilitate the following regioselective addition of activated oxygen to Aci-reductone ligand and the reactions leading to formation of methionine. Here for the first time we demonstrate the different structures organization of complexes of iron in aqueous and hydrocarbon medium.

First, we received UV-spectrum data, testified in the favor of the complex formation between $Fe(acac)_3$ and 18C6. In the next Fig. 4.13, the spectrums of solutions of $Fe(acac)_3$ (1) and mixture $\{Fe(acac)_3$+18C6$\}$ (2) in various solvents are presented.

As one can see in Fig. 4.13a, at the addition of the 18C6 solution (in $CHCl_3$) to the $Fe(acac)_3$ solution (in $CHCl_3$) (1:1) an increase in maximum of absorption band in spectrum for acetylacetonate-ion $(acac)^{-}$ in complex with iron, broadening of the spectrum and a bathochromic shift of the absorption maximum from $\lambda \sim 285$ nm to $\lambda = 289$ nm take place. The similar changes in the intensity of the absorption band and shift of the absorption band are characteristic for narrow, crown unseparated ion-pairs [9]. Earlier similar changes in the UV – absorption band of $Co^{II}(acac)_2$

solution we observed in the case of the coordination of macrocyclic polyether 18C6 with $Co^{II}(acac)_2$ [9]. The formation of a complex between $Fe(acac)_3$ and 18C6 occurs at preservation of acac ligand in internal coordination sphere of Fe^{III} ion because at the another case the short-wave shift of the absorption band should be accompanied by a significant increase in the absorption of the solution at $\lambda = 275$ nm, which correspond to the absorption maximum of acetyl acetone [9]. It is known that Fe^{II} and Fe^{III} halogens form complexes with crown-ethers of variable composition (1:1, 1:2, 2:1) and structure dependent on type of crown-ether and solvent [20]. It is known that $Fe(acac)_3$ forms labile OSCs (Outer Sphere Complexes) with $CHCl_3$ due to H-bonds [21].

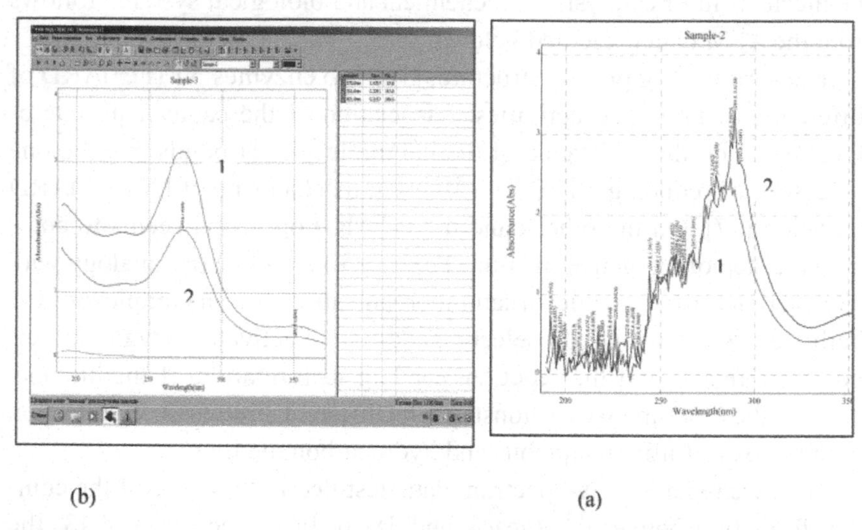

FIGURE 4.13 Absorption spectra of $CHCl_3$ solutions: of $Fe(acac)_3$ (1), mixture $\{Fe(acac)_3 + 18C6\}$(1:1) (2): a – in $CHCl_3$, b – in H_2O, 20°C.

However, in an aqueous medium the view of UV-spectrum is changing (Fig. 4.12), b: an decrease in absorption maximum of acetylacetonate ion $(acac)^-$ (in $Fe(acac)_3$) at the addition of a solution of 18C6 to the $Fe(acac)_3$ solution (1:1). Possibly, in this case inner-sphere coordination of 18C6 cannot be excluded.

In an aqueous medium the formation of supramolecular structures of generalized formula $Fe^{III}{}_x(acac)_y 18C6\ {}_m(H_2O)_n$ is quite probable.

In the Fig. 4.14, three-dimensional (a) and two-dimensional (b) AFM image of the structures on the basis of iron complex with 18C6 FeIIIx(acac) y18C6 m(H2O)n, formed at putting a uterine solution on a hydrophobic surface of modified silicone are presented. It is visible that the generated structures are organized in certain way forming structures resembling the shape of tubule micro fiber cavity (Fig. 4.15c). The heights of particles are about 3–4 nm. In control experiments it was shown that for similar complexes of nickel NiII(acac)2·18C6·(H2O)n (as well as complexes Ni2(OAc)3(acac)·MP·2H2O) this structures organization is not observed. It was established that these iron constructions are not formed in the absence of the aqueous environment. Earlier in our works we showed the participation of H2O molecules in mechanism of FeIII, IIx(acac)y18C6 m(H2O)n transformation by analogy Dke1 action, and also the increase in catalytic activity of iron complexes (FeIIIx(acac)y18C6 m(H2O)n, FeIIx(acac)y18C6 m(H2O)n and FeIIxL1y(L1ox)z(18C6)n(H2O)m) in the ethyl benzene oxidation in the presence of small amounts of water [6, 9]. In [22] it was found that the possibility of decomposition of the b-diketone by analogy with Fe-ARD' action increases in aquatic environment. That apparently is consistent with data, obtained in our previous works [6, 9].

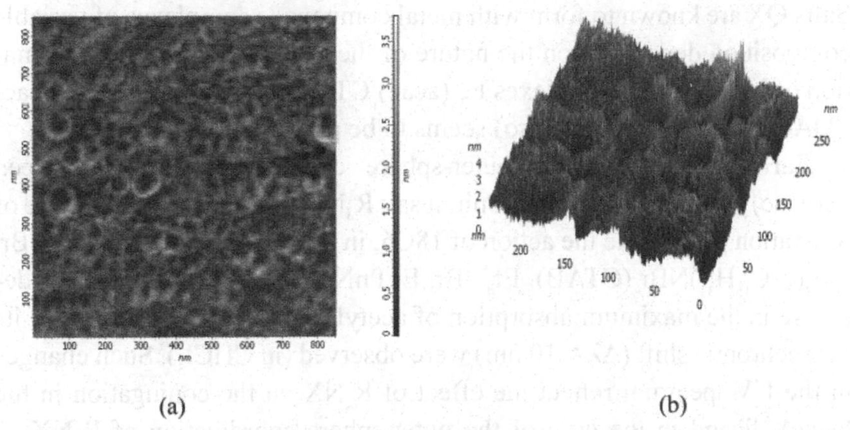

FIGURE 4.14 The AFM two- (a) and three-dimensional (b) image of nanoparticles on the basis $Fe_x(acac)_y 18C6\ _m(H_2O)_n$ formed on the surface of modified silicone.

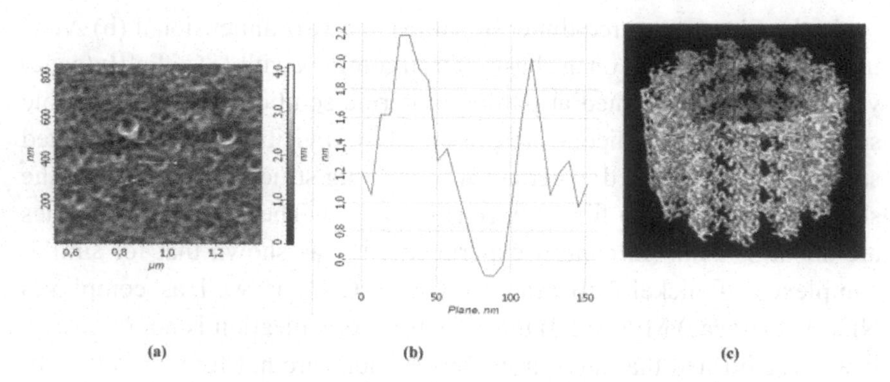

FIGURE 4.15 The AFM two-dimensional image (a) of nanoparticles on the basis $Fe_x(acac)_y18C6\ _m(H_2O)_n$ formed on the hydrophobic surface of modified silicone. The section of a circular shape with fixed length and orientation is about 50–80 nm (b), (c) The structure of the cell microtubules.

As another example we researched the possibility of the supramolecular nanostructures formation on the basis of $Fe_x(acac)_yCTAB_m$ at putting of solutions of $Fe_x(acac)_yCTAB_m$ in $CHCl_3$ (Fig. 4.16) or H_2O (Fig. 4.17) on the hydrophobic surface of modified silicon ($CTAB=Me_3(n\text{-}C_{16}H_{33})NBr$). We used CTAB concentration in the 5–10 times less than $Fe(acac)_3$ concentration to reduce the probability of formation of micelles in water. But formation of spherical micelles at these conditions cannot be excluded. Salts QX are known to form with metal compounds complexes of variable composition depending on the nature of the solvent [5, 9]. So the formation of heteroligand complexes $Fe_x(acac)_yCTAB_m(CHCl_3)_p$ (and $Fe^{II}_x(acac)_y(OAc)_z(CTAB)_n(CHCl_3)_q$ also) seems to be probable [21].

Earlier we established outer-sphere complex formation between $Fe(acac)_3$ and quaternary ammonium salt R_4NBr with different structure of R – cation [9]. Unlike the action of 18C6, in the presence of salts Me_4NBr, $Me_3(n\text{-}C_{16}H_{33})NBr$ (CTAB), Et_4NBr, Et_3PhNCl, Bu_4NI and Bu_4NBr, a decrease in the maximum absorption of acetylacetonate ion $(acac)^-$, and its bathochromic shift ($\Delta\lambda \approx 10$ nm) were observed (in $CHCl_3$). Such changes in the UV-spectrum reflect the effect of R_4NX on the conjugation in the $(acac)^-$ ligand in the case of the outer-sphere coordination of R_4NX. A change in the conjugation in the chelate ring of the acetylacetonate complex could be due to the involvement of oxygen atoms of the acac ligand in the formation of covalent bonds with the nitrogen atom or hydrogen bonds with CH groups of alkyl substituents [9].

In the Fig. 4.16, three-dimensional (a) and two-dimensional (b, d) AFM image of the structures on the basis of iron complex – $Fe_x(acac)_yCTAB_m(CHCl_3)_p$, formed at putting a uterine solution on a surface of modified silicone are presented.

As one can see, these nanostructures are similar to macrostructures, observed previously, but with less explicit structures, presented in Figs. 4.14 and 4.15, and which resemble the shape of tubule micro fiber cavity (Fig. 4.15c). The height of particles on the basis of $Fe_x(acac)_yCTAB_m(CHCl_3)p$ is about 7–8 nm (Fig. 4.16c), that more appropriate parameter for particle with 18C6 $Fe_x(acac)_y18C6\ _m(H_2O)_n$. We showed that complexes of nickel $Ni^{II}(acac)_2 \cdot CTAB$ (1:1) (in $CHCl_3$) do not form similar structures.

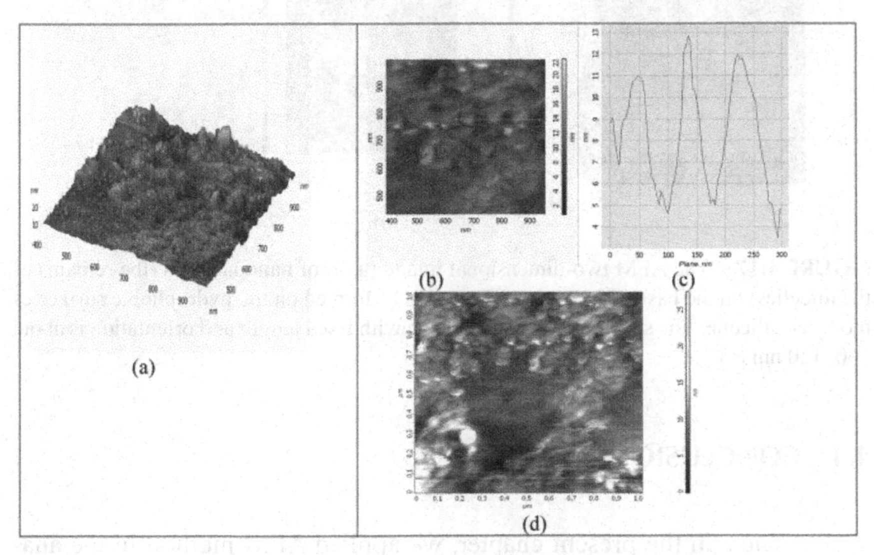

FIGURE 4.16 The AFM three- (a) two-dimensional image (b, d) of nanoparticles on the basis $Fe_x(acac)_yCTAB_m(CHCl_3)_p$ formed on the hydrophobic surface of modified silicone. The section of a circular shape with fixed length and orientation is about 50–70 nm (c).

In the Fig. 4.17, two-dimensional (a–c) AFM image of the structures on the basis of iron complex with CTAB: $Fe_x(acac)_yCTAB_p(H_2O)_q$, – formed at putting a uterine water solution on a hydrophobic surface of modified silicone, are presented. In this case, we observed apparently phenomenon of the particles, which are the remnants of the micelles formed. Obviously, the spherical micelles, based on $Fe(acac)_3$ and CTAB, on a hydrophobic surface are extremely unstable and decompose rapidly. The remains of the micelles have circular shaped structures, different from that observed in Figs. 4.15 and 4.16. The heights of particles on the basis of $Fe_x(acac)_yCTAB_p(H_2O)_q$ (Fig. 4.17) are about 12–13 nm that are greater than in the case of particles presented in Figs. 4.15 and 4.16. The particles in Fig. 4.17 resemble micelles in sizes (particle sizes in the XY plane are 100–120 nm (Fig. 4.17c)) and also like micelles, probably due to the rapid evaporation of water needed for their existence, are very unstable and destroyed rapidly (even during measurements).

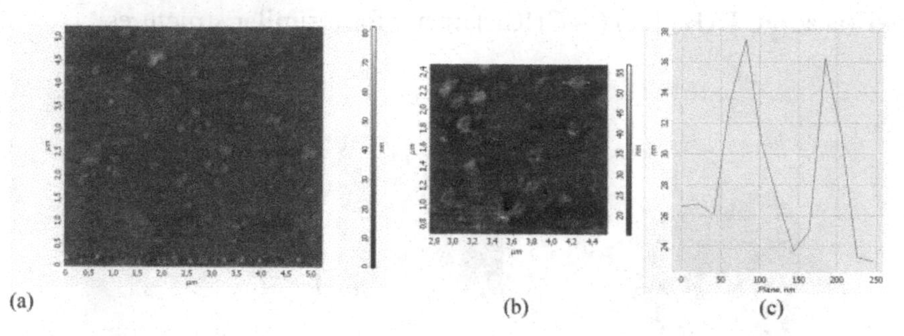

(a) (b) (c)

FIGURE 4.17 The AFM two-dimensional image (a, b) of nanoparticles (the remains of the micelles) on the base of $Fe_x(acac)_yCTAB_p(H_2O)_q$ formed on the hydrophobic surface of modified silicone. The section of a circular shape with fixed length and orientation is about 100–120 nm (c).

4.4 CONCLUSION

1. Thus, in the present chapter, we applied AFM method in the analytical purposes to research the possibility of the formation of supramolecular structures on basis of heterobimetallic heteroligand triple complexes $Ni^{II}(acac)_2 \cdot MSt \cdot PhOH$ (M=NaSt, LiSt), heteroligand complexes $Ni_{2(OAc)3(acac)} \cdot MP \cdot 2H_2O$ (L^2 = MP), $Fe_x(acac)_y18C6_m(H_2O)_n$ and $Fe_x(acac)_yCTAB_m(CHCl_3)_p$ with the assistance of intermolecular H-bonds.

We have shown that the self-assembly driven growth seems to be due to H-bonding of structures on the basis of $Ni^{II}(acac)_2 \cdot MSt \cdot PhOH$ with a surface of modified silicone, and further formation supramolecular nanostructures $\{Ni^{II}(acac)_2 \cdot MSt \cdot PhOH\}_n$ due to directional intermolecular (phenol–carboxylate) H-bonds, and, possibly, other noncovalent interactions (van Der Waals-attractions and π-bonding).

These data support the very probable supramolecular structures appearance on the basis of heterobimetallic heteroligand triple complexes $Ni^{II}(acac)_2 \cdot MSt \cdot PhOH$ in the course of the ethyl benzene oxidation with dioxygen, catalyzed by three-component catalytic system $\{Ni^{II}(acac)_2 + MSt + PhOH\}$ and this can be one of the explanations of the high values of conversion of the ethyl benzene oxidation into α-phenyl ethyl hydro peroxide at selectivity S_{PEH} preservation at level not below S_{PEH}=90 % in this process. The higher effectivity of heterobimetallic heteroligand complexes $Ni^{II}(acac)_2 \cdot NaSt \cdot PhOH$, including metalloligand NaSt, as selective catalysts in comparison with $Ni^{II}(acac)_2 \cdot LiSt \cdot PhOH$, seems to be due to formation of more stable supramolecular structures $\{Ni^{II}(acac)_2 \cdot NaSt \cdot PhOH\}_n$ during hydrocarbon oxidation.

2. The experimental data, obtained by using AFM method, indicate high probability of supramolecular structures formation on the basis of complex $Ni_{2(OAc)3(acac)\cdot MP\cdot 2H2}O$ with the assistance of intermolecular H-bonds in the real systems, namely, in the processes of alkylarens oxidation. H-bonding seems to be one of the factors, responsible for the stability of real catalysts – heteroligand complexes $Ni_{2(OAc)3(acac)\cdot MP\cdot 2H2}O$ $(Ni_x(acac)_y(OAc)_z(L^2)_n(H_2O)_m)$ in the course of alkylarens (ethyl benzene, cumene) oxidation by dioxygen into hydro peroxide (intermediates in the large-scale production of important monomers) in the presence of catalytic systems $\{Ni^{II}(acac)_2 + L^2\}$.

The data received can be useful in treatment of biological effects in $Ni^{II}(Fe^{II})$ARD enzymes operation.

The formation of multidimensional forms (in the case of Ni^{II}ARD) may be one way of controlling $Ni^{II}(Fe^{II})$ARD activity [17]. Specific structural organization of iron complexes may facilitate the first step in $Fe^{II}ARD$ operation: O_2 activation and following regioselective addition of activated oxygen to Acireductone ligand (unlike mechanism of regioselective addition of nonactivated O_2 to Acireductone ligand in the case of $Ni^{II}ARD$), and reactions leading to formation of methionine.

At the same time it is necessary to mean that important function of $Ni^{II}ARD$ in cells is established now. Namely, carbon monoxide, CO, is formed as a result of action of nickel-containing dioxygenase $Ni^{II}ARD$. It was established, that CO is a representative of the new class of neural messengers, and seems to be a signal transducer like nitrogen oxide, NO [15, 17].

KEYWRODS

- **AFM method**
- **binary and triple catalytic systems**
- **mechanisms of catalysis**
- **nickel and iron heteroligand complexes**
- **supramolecular chemistry**
- **supramolecular nanostructures**

REFERENCES

1. St. Leninger, B. Olenyuk, P.J. Stang, Self-Assembly of Discrete Cyclic Nanostructures Mediated by Transition Metals, Chem. Rev., 100(3), 853–908, 2000.
2. P.J. Stang, B. Olenyuk, Self-Assembly, Symmetry, and Molecular Architecture: Coordination as the Motif in the Rational Design of Supramolecular Metallacyclic Polygons and Polyhedra, Acc. Chem. Res., 30(12), 502–518, 1997.
3. C.M. Drain, A.I. VarottoRadivojevic, Self-Organized Porphyrinic Materials, Chem. Rev., 109(5), 1630–1658, 2009.
4. I. Beletskaya, V.S. Tyurin, A. Yu. Tsivadze, R. Guilard, Ch. Stem, Supramolecular Chemistry of Metalloporphyrins, Chem. Rev., 109(5), 1659–1713, 2009.
5. L.I. Matienko, Solution of the problem of selective oxidation of alkylarenes by molecular oxygen to corresponding hydroperoxides. Catalysis initiated by Ni(II), Co(II), and Fe(III) complexes activated by additives of electron-donor mono- or multidentate

extraligands, In: Reactions and Properties of Monomers and Polymers, A. D'Amore and G. Zaikov (Eds.), Chapter 2, New York: Nova Science Publ. Inc., 2007, p.21–41.

6. L.I. Matienko, L.A. Mosolova, The Modeling of Catalytic Activity of complexes Fe(II, III)$(acac)_n$ with R_4NBr or 18-crown-6 in the Ethylbenzene Oxidation by Dioxygen in the presence of small amounts of H_2O. Oxid. Commun., 33(4), 830–844, 2010.
7. K. Weissermel, H.-J. Arpe, Industrial Organic Chemistry, 3nd ed., transl. by Lindley C.R.. New York: VCH, 1997.
8. A.K. Suresh, M.M. Sharma, T. Sridhar, Industrial Hydrocarbon Oxidation, *Ind. Eng. Chem. Res*. 39(11), 3958–3969, 2000.
9. L.I. Matienko, L.A. Mosolova, G.E. Zaikov, Selective Catalytic Hydrocarbons Oxidation. New Perspectives, New York: Nova Science Publ. Inc., USA, 2010, 150 P.
10. Y.J. Park, J.W. Ziller, A.S. Borovik, The effect of Redox-Inactive Metal Ions on the Activation of Dioxygen: Isolation and Characterization of Heterobimetallic Complex Containing a Mn^{III}-(μ-OH)-Ca^{II} Core, J. Am. Chem. Soc., 133(24), 9258–9261, 2011.
11. M. Dubey, R.R. Koner, M. Ray, Sodium and Potassium Ion Directed Self-assembled Multinuclear Assembly of Divalent Nickel or Copper and L-Leucine Derived Ligand, *Inorg. Chem.*, 48(19), 9294–9302, 2009.
12. E.V. Basiuk, V.V. Basiuk, J. Gomez-Lara, R.A. Toscano, A bridged high-spin complex bis- [Ni(II)(rac-5,5,7,12,12,14-hexamethyl-1,4,8,11-tetraazacyclotetradecane)]-2,5-pyridinedicaboxylate diperchlorate monohydrate, *J. Incl. Phenom. Macrocycl. Chem.* 38(1), 45–56, 2000.
13. P.Mukherjee, M.G.B. Drew, C.J. Gómez-Garcia, A. Ghosh, (Ni_2), (Ni_3), and $(Ni_2 + Ni_3)$: A Unique Example of Isolated and Cocrystallized Ni_2 and Ni_3 Complexes, *Inorg. Chem.*, 48(11), 4817–4825, 2009.
14. Ludmila Matienko, Vladimir Binyukov, Larisa Mosolova and Gennady Zaikov, The selective ethylbenzene oxidation by dioxygen into α-phenyl ethyl hydroperoxide, catalyzed with triple catalytic system $\{Ni^{II}(acac)_2$+NaSt(LiSt)+PhOH}. Formation of nanostructures $\{Ni^{II}(acac)_2\cdot NaSt\cdot (PhOH)\}_n$ with assistance of intermolecular H-bonds. Polymers Research Journal, 5(4), 423–431, 2011.
15. Y. Dai, Th.C. Pochapsky, R.H. Abeles, Mechanistic Studies of Two Dioxygenases in the Methionine Salvage Pathway of *Klebsiella pneumoniae*, Biochemistry, 40(21), 6379–6385, 2001.
16. G.D. Straganz, B. Nidetzky, Reaction Coordinate Analysis for *β*-Diketone Cleavage by the Non-Heme Fe^{2+}-Dependent Dioxygenase Dke 1, J. Am. Chem. Soc., 127(35), 12306–12315, 2005.
17. S.C. Chai, T. Ju, M. Dang, R.B. Goldsmith, M.J. Maroney, Th.C. Pochapsky, Biochemistry, 47(8), 2428–2435, 2008.
18. B. Gopal., L.L. Madan, S.F. Betz, and A.A. Kossiakoff, The Crystal Structure of a Quercetin 2,3-Dioxygenase from *Bacillus subtilis* Suggests Modulation of Enzyme Activity by a Change in the Metal Ion at the Active Site(s), Biochemistry, 44(1), 193–205, 2005.
19. L.I. Matienko, L.A. Mosolova, V.I. Binyukov, E.M. Mil, G.E. Zaikov "The new approach to research of mechanism catalysis with nickel complexes in alkylarens oxidation" "Polymer Yearbook" 2011. N.-Y.: Nova Science Publishers, 2012, 221–230.

20. V.K. Belskii, B.M. Buleachev, Structurally chemical aspects of complex forming in systems metal halide – a macrocyclic polyether, *Uspekhi khimii*, 68(2), 136–147, 1999 (*in Russian*).
21. V.M. Nekipelov, K.I. Zamaraev, Outer-sphere coordination of organic molecules to electric neutral metal complexes, Coord. Chem. Revs., 61(1), 185–203, 1985.
22. C.J. Allpress, K. Grubel, E. Szajna-Fuller, A.M. Arif, and L.M. Berreau, Regioselective Aliphatic Carbon-Carbon Bond Cleavage by Model System of Relevance to Iron-Contaning Acireductone Dioxygenase, J. Am. Chem. Soc., 135(2), 659–668, 2013.

CHAPTER 5

SYNTHESIS OF SYNTHETIC MINERAL-BASED ALLOYS LIQUATION PHENOMENA OF DIFFERENTIATION

A. M. IGNATOVA and M. N. IGNATOV

CONTENTS

5.1 Introduction 190
Keywords 197
References 197

5.1 INTRODUCTION

Natural stone – is a unique material, which is the standard combination of durability, reliability and beauty. Therefore, natural stone is still popular, along with modern synthetic materials, especially at a time when important esthetic qualities.

A variety of patterns and textures of natural stone provides petrochemical processes that occur in the magma at the stage of formation of rocks. To form the texture of the stone the most important process is the separation of the melt to separate liquid phases. These phases differ in density, composition, color, shape, mineral aggregates, etc.

The role of magmas in the formation of bundles of macro and microstructure of the rocks has been evaluated, recently. In chemistry, the existence of two immiscible liquids of different densities has long been known, the simplest example of such a system of "oil-water." At the end of the XIX century. The view emerged that immiscibility may be present in the molten silicate magma, this phenomenon is called "phase separation" or "liquation differentiation" [1]. Study of this phenomenon and its influence on the formation of the rocks studied by well-known foreign and domestic scholars like Levinson-Lessing, Bowen and Greig [2, 3]. The main directions of the discussion around the phenomenon of phase separation, there were two: whether it exists and if there is any rock may result from this process. Much less attention is paid to the crystallization of silicate melt stratified rocks. It was believed that each of the liquid phases crystallize separately, suggesting that these processes occur in parallel [4].

With the development of research techniques, it was shown that phase separation is present in some granites, basalts, as well as in lunar rocks [5]. It was found that the effect of stratification is present in virtually all glass and many other synthetic silicate melts. The phenomenon of phase separation is most effective in the study of synthetic materials. Of all the materials that separate into two liquid phases, synthetic mineral alloys are most similar to the rocks [6].

Synthetic mineral alloys (siminals) – a type of petrurgical materials obtained by crystallization of molten rock, or various waste products of basic and ultra basic character. Siminals – it is also a material obtained by the relaxation of physical and chemical processes associated with the cooling of the melt produced by oxide mixture with a high concentration of silica.

Siminals structure is a combination of crystalline aggregates, ranging in size from 100 nm to 800 microns, and an amorphous glassy phase with nanoscale nucleation.

Earlier phase separation siminals not investigated because it was believed that the phenomenon of immiscibility exists only at high temperatures, and hence, by the time of crystallization, even if there was segregation, the melt becomes homogeneous state. We believe that the liquid phase in the melt siminals certainly interact, as well moistened with each other. The study of the crystallization of silicate melts from the separated state, taking into account the interaction between the phases is an urgent task. The solution to this problem will manage the processes of phase separation in order to achieve certain properties siminals, including esthetic. The purpose of this study is to examine the interaction of fiber in the melt phase during crystallization siminals.

The phenomenon of phase separation in the melts siminals was first discovered in the industrial manufacture of products from them for stone casting technology. Watching the process, the engineers found that by pouring superheated melt products are highly fragile. On the cleaved products such visible banded structure. The shape and distribution of contrasting bands of repeating the direction of flow of melt in the mold. This texture is called "tree." Since the phenomenon of phase separation led to the marriage of products, the only focus of his study was to identify the conditions under which it cannot arise. Once these conditions are found in practice, research on this matter have ceased.

In the works of Fenner [7] notes that melt into fibers when the system loses its balance. The melt can lose your balance when lowering the temperature or the temperature drops and the simultaneous crystallization. Therefore, segregation can be stable and metastable [8]. Recognized as a stable one that occurs before the liquidus and the metastable one which occurs in the interval between the liquidus and solidus. In practice, most siminals production is a metastable phase separation.

In the works of American Scientists Rutherford and Hoffman [9, 10], noted that the allocation of droplets of one liquid in another, a further change in composition of these fluids depends on the diffusion of elements in both melts and the surface phase sections. This process is similar to the processes of nucleation and crystallization of solid solutions from the melt, but the rate is much higher, as occurs in fluid.

Visually segregation in siminals has banded texture on the macro – and micro – level (Fig. 5.1), since the melt siminals dynamically positioned (poured into a mold). Liquation siminals texture can be described as a mass matrix are areas with very different composition and structure.

In 1987, E. Roeder [11], a famous scholar of lunar rocks, explained a clear selectivity, because of the growing crystal is a two-phase melt inclusions in the form captures that fluid, which strongly differs from it in composition, since the components of the second liquid consumed with growth.

Therefore, the phase separation in siminals is not a process parallel to the crystallization of two liquid phases, and the process of combating a specific mineral phases for the "Leadership," a kind of "expansion."

Since the beginning of deposition of droplets of one phase relative to the other, immediately begins a rapid exchange of elements. One phase is the one whose free energy at the surface above begins to actively grow at the expense of pumping from the other siblings, capturing some parts of "donor" phase.

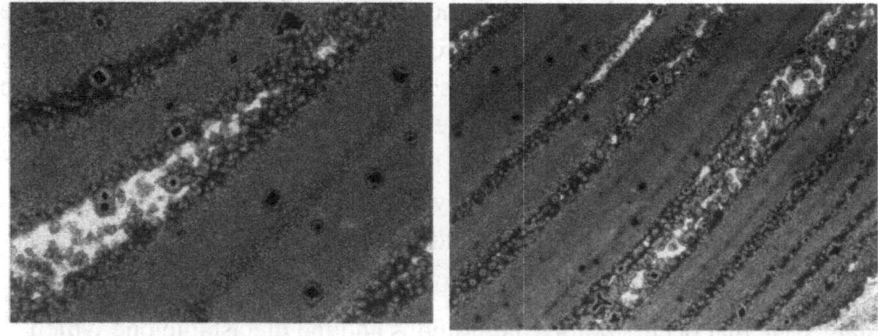

FIGURE 5.1 Siminals structure with the addition of chromite in the charge, an optical microscope, x100 (nicols crossed).

However, the images obtained by optical microscopy revealed the presence of not two components, and three. As already mentioned, the phase separation in the metastable siminals has character, that is, at the time of separation in the melt is already present, and even crystal nucleation in the initial stages of growth. Nucleation and crystal nuclei are formed by the phases, the melting point of which the most high. Previous studies of phase composition siminals with phase separation [12] showed that, in their present diopside, augite, aegirine, and quartz (Fig. 5.2). Moreover, the composition of diopside $(Na_5Ca_5)(Cr_5\ Mg_5)Si\ (Al)\ _2O_6$, the concentration of chromium in siminals low, about 5–6%, so it is obvious that the first nucleation will be the nucleation of diopside. However, their growth will be limited due to the low concentration of chromium in the melt. Therefore, nucleation of diopside can develop into a small chip located at a large distance from each other.

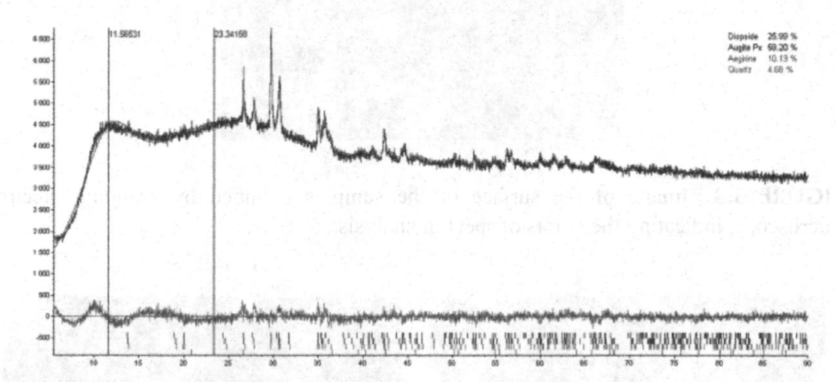

FIGURE 5.2 Diffractogram of the sample siminals a fixed state of phase separation.

TABLE 5.1 The composition of the elements at the points of the spectral analysis.

The point of the analysis	The content of the element. %										
	O	Mg	Al	Si	Ti	V	Cr	Fe	Na	K	Ca
spectrum 1	40.63	5.33	6.12	27.08	0.89	—	0.24	10.32	0.86	0.28	8.36
spectrum 4	13.82	4.28	3.43	0.34	0.45	0.4	54.16	23.14	—	—	—
spectrum 5	46.57	5.55	5.79	25.52	0.78	—	—	7.75	1.01	0.19	6.84

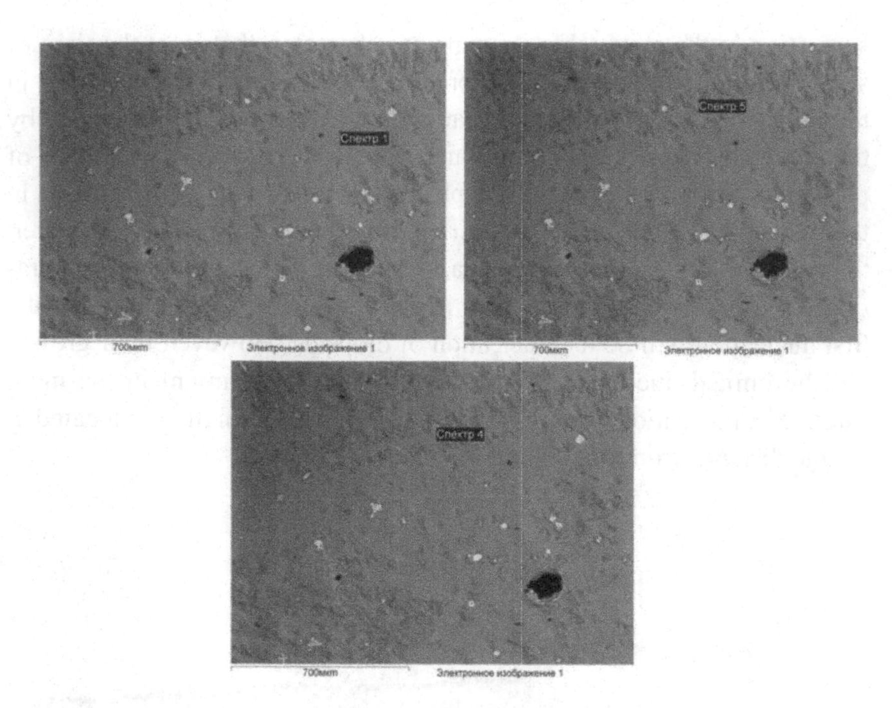

FIGURE 5.3 Image of the surface of the samples obtained by scanning electron microscopy, indicating the points of spectral analysis.

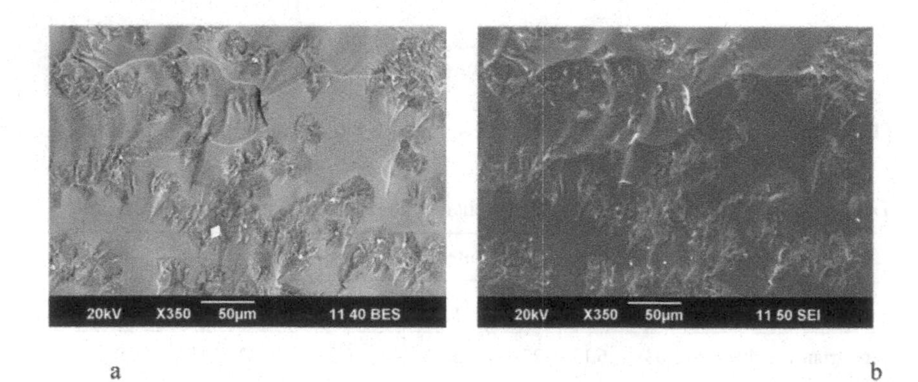

a b

FIGURE 5.4 Image of the surface area of a sample surface at different shooting modes for electron microscopy: a – a chemical contrast, b – topographical contrast.

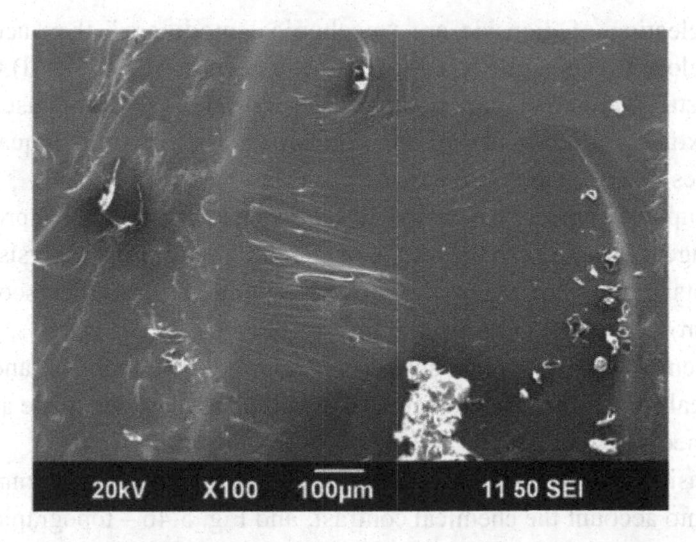

FIGURE 5.5 The site of fracture on the surface of the hardened area in the amorphous-glassy state.

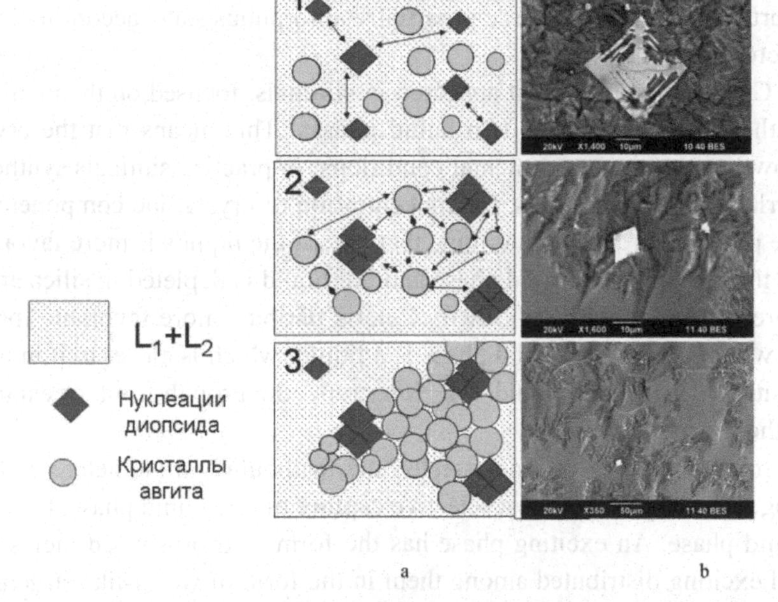

FIGURE 5.6 The order of crystallization siminals of the two-phase fluid: a – scheme, b – electron microscopy images illustrating the pattern.

Nucleation of diopside are "mechanical modifiers," they acquire a phase close to the augite (Ca, Na) (Mg, Fe^{2+}, Fe^{3+}, Al^{3+}) [$(Si, Al)_2O_6$]. As the crystallization of augite captures a small drop of "donor" phase, which is a mixture of quartz and aegirine, aegirine part of a mixture of quartz and solidifies in amorphous glassy state.

In support of the above-described sequence of crystallization processes are images (Fig. 5.3) and the results of spectral microprobe analysis (Table 5.1) obtained by electron microscopy. Scanning electron microscopy and microprobe analysis performed at the facility JSM-6390LV.

When taking the sample surface in the modes of chemical and topographical contrasts revealed that the theoretical assumption made above is confirmed.

Consider the images of the same plot. In Fig. 5.4, and the image that takes into account the chemical contrast, and Fig. 5.4b – topography. Lots of diopside, which are in stark contrast to the chemical composition, almost not detectable visually on the topography. From this we can conclude that they are in the surrounding augite formations. Augite is strongly distinguished from the height of the relief of that part which is a mixture of quartz and aegirine (it hardens partially amorphous state, according to the fracture, Fig. 5.5).

Crystalline phases that appeared in siminals, focused on the meniscus of the interface between two liquid phases. This means that the crystal growth under the experimental conditions, in practice siminals synthesis, carried out the two fluids. He concentration of crystalline components in one phase can reliably determine that one of the liquids is more favorable for the growth of the solid phase and the liquid is depleted in silica and is more consumed with its growth. Ceteris paribus, more favorable for the growth of a particular solid phase is a liquid, which is closer to it in composition. The second liquid will either repel the growth front, or captured in the form of inclusions.

It turns out that, in a metastable state, without regard to hetero-and homogeneous nucleation, is a selective capture of one liquid phase, the other liquid phase. An exciting phase has the form of drop-shaped inclusions, and exciting distributed among them in the form of worm-like structures. We can say that the relaxation processes in the crystallization of the melt phase separation take place successfully in some phases, but in general the complete relaxation occurs.

The successive stages of crystallization of the two-phase liquid melt siminals we presented graphically in Fig. 5.6, and the most indicative and reinforced our view of electron microscopy images (Fig. 5.6b).

Thus, thanks to the research found that in general the whole process of structure formation of the two-phase melt siminals can be characterized as a process catalyzed exchange of items between the two phases on the basis of the basicity in the crystallization conditions. As a result, one of the other fluids trapped as inclusions, there are areas with a chaotic distribution of components and atypical morphology, which follows the shape of one of the two-phase melt of liquid phases. This allows further, using the installed feature, to manage the process of phase separation of silicate melts, predicting the composition of the liquid phases, the nature of their interaction at different stages, and hence to achieve a certain color contrast of these phases, the amount of layers to separate them in the solid state and thus does not mimic the natural beauty and harmony of stone and reach her, driving petrochemical reactions in the melt.

KEYWORDS

- **high concentration of silica**
- **liquation differentiation**
- **macro and microstructure of the rocks**
- **mineral-based alloys**
- **phase separation**
- **synthetic mineral alloys**

REFERENCES

1. Becker, G. F., Some queries on rock differentiation, Am. Sci, J., 3, 21–40, 1897.
2. Levinson-Lessing, F. Ju., Petrografija, 5 izd., Magma L., 1940
3. P. I. Lebedev. Akademik, F. Ju. Levinson-Lessing kak teoretik petrografii Izd-vo Akademii Nauk SSSR, M.-L., 1947 g.

4. Redder Je. Likvacija silikatnyh magm. Jevoljucija izverzhennyh porod. Pod red, X. Jodera. M.: Mir. 1983. S. 24–67.
5. Roedder, E., Weibien, P. W. Lunar petrology of silicate melt inclusions, Apollo 11 rocks. Proceedings of the Apollo 11 lunar science conference. V. 1 (Mineralogy and petrology). Pergamon Press, 1970. P. 801–837.
6. Ignatova, A. M. Priroda likvacionnyh javlenij V sinteticheskih rasplavah kamennogo lit'ja. Uspehi sovremennogo estestvoznanija. № 8, 2010. 22–23.
7. Fenner, S. N., The crystallization of basalts. Am. J. Sci., 18, 225–253, 1929.
8. Martin, B., Kushiro, I. Immiscibility synthesis as an indicator of cooling rates of basalts.Volcanol. Geothermal Res. 1991. 45. P. 289–310.
9. Hess, P. C., Rutherford, M. J., Guillemette, R. N., Reyrson, F. J., Tuchfeld, H. A., Residual products of fractional crystallization of lunar magmas: An experimental study, Proc. Sixth Lunar Sci. Conf., Geochim. Cosmo. Acta Suppl., 6, 1–895–909.
10. Hoffmann, A. W., Diffusion of Ca and Sr in a basalt system, Carnegie Institution of Washington Year Book 74, for 1974–1975, 183–189, 1975.
11. Redder Edwin, The role of liquid immiscibility in igneous petrogenesis: a discussion, J. Geology, 64, 84–88. 1956.
12. Ignatova, A. M., Nikolaev, M. M., Hanov, A. M., Chernov V. P. Issledovanie i razrabotka osnovnyh pravil upravlenija strukturnym mirom silikatov i tehnologij poluchenija steklokristallicheskogo i sljudokristallicheskogo kamennogo lit'ja. Materialy XLVII MNSK "Student i nauchno-tehnicheskij progress," sekcija: "Himija" Novosibirsk: NGU, 2009, 161.

CHAPTER 6

RESTRUCTURING OF SYNTHETIC MINERAL ALLOYS UNDER IMPACT

A. M. IGNATOVA and M. N. IGNATOV

CONTENTS

6.1 Introduction 200
6.2 Aim and Object of Study 201
6.3 Research Methods and Results 202
6.4 Discussion of the Results 207
6.5 Conclusions 209
Keywords 210
References 210

6.1 INTRODUCTION

It is known that many rocks are formed under intense impact effects (e.g., impact of meteorites). Deposits of minerals, including rare metals and nonmetal structures, are found in impact craters (astrobleme) [1–3]. Additionally, impact metamorphism forms polymorphs of certain minerals that cannot be formed under other conditions, such as the modification of quartz to form coesite and stishovite [4].

It has been established experimentally that some impact breeds can be synthesized. Such synthesis is possible under the impact of a small number of nonaltered rock hits under the condition that the force of the impact and volume of rock is well correlated with the force of meteorite impact and volume of mountain formation [5, 6].

Clearly, impact effects change the structure of synthetic materials, not only naturally occurring materials. Consequently, impact can be considered a method of synthesis, not only a destructive phenomenon.

The principles of impact metamorphism are used, in part, for mechanical activation [7]. Indeed, the mechanical activation of powder materials can be used to obtain a variety of materials with very fine structure. However, the structural changes induced by mechanical activation are different from those induced by impact metamorphism. Shock metamorphism occurs due to the effects of impact on a monolithic object, while mechanical activation acts on powder materials. Thus, it is important to study the structural changes in materials in the solid and monolithic state under shock impact.

It is known that under shock impact, impact energy is transferred; this phenomenon is called energy dissipation. In this process, which has several different stages, the impact energy is converted to heat [8].

Different materials dissipate energy differently, depending on their structure. A good example is how materials behave when hit by a bullet from a gun. If a material is destroyed by a bullet that stops moving within the material, then the kinetic energy of the bullet is spent on the destruction and changes in the material. Conversely, if the bullet continues to move, and the material is damaged slightly, then such a material is not prone to dissipation. The more a material able to dissipate energy, the more its structure changes under shock impact. Rocks are materials with a high capacity for dissipation, as confirmed by the phenomenon of impact metamorphism described previously. Materials with structures similar to those of natural mineral formations will have a similar capacity for dissipation.

Rock materials such as glasses and ceramics are often used to produce armor. Armor is necessary not only for protection from firearms but also to protect turbine engine blades, for example, in the case of emergency isolation.

However, the materials with greatest similarity to natural mineral formations are synthetic mineral alloys (siminals). The term "siminals" is new, and it is intended to replace the outdated term "casting a stone."

Siminals are materials obtained by remelting one or more varieties of rocks or man-made mineral formations. Their structure and composition are similar to those of mafic and ultramafic igneous rocks. The structure contains a siminal amorphous phase (2 to 30%) and two or more mineral phases. Siminals are not glass-crystalline materials because they are not crystallized using catalysts [9]. Siminals contain an average of 50% SiO_2; therefore, siminal melts often split into two liquid phases. The structure of siminals represents a special case of a condensed medium.

6.2 AIM AND OBJECT OF STUDY

The aim of this study was to study structural changes in siminals, specifically raw hornblendite materials, under shock impact.

The chemical composition of the samples, as presented in Table 6.1, was obtained by X-ray spectral fluorescence (CPM-18, EDX 900HS).

TABLE 6.1 Composition of samples studied.

The content of components, %								
MgO	Al_2O_3	SiO_2	K_2O	CaO	TiO_2	Cr	MnO	Fe_2O_3
9.91	10.82	49.6	0.22	10.33	1.38	0.54	0.16	14.02

6.3 RESEARCH METHODS AND RESULTS

High-speed penetration tests were conducted by subjecting samples to shock cylindrical punching at speeds of 125, 600 and 800 m/s [13]. At the moment of impact, the temperature distribution over the punch pattern on the backside of the samples was recorded using an infrared camera.

Changes in the structure of the samples were evaluated using X-ray diffraction and scanning electron microscopy, with associated spectral microprobe analysis. The fragments of the destroyed samples were studied.

X-ray diffraction was conducted using an ESR70–30 DX/2 diffractometer. Scanning electron microscopy and microprobe analysis were performed on a JSM-6390LV.

Radiographs of the samples after run off at a speed of 125 m/s are shown in Fig. 6.1. The spectrogram revealed that the main phase of the samples was clinopyroxene (diopside-augite composition), with olivine, quartz, and traces of chromite impurities also observed. The greatest interest in terms of the changes in the structure lies in the intensity of the peak at 2.96, which corresponds to one of the shock-induced polymorphs of quartz – stishovite.

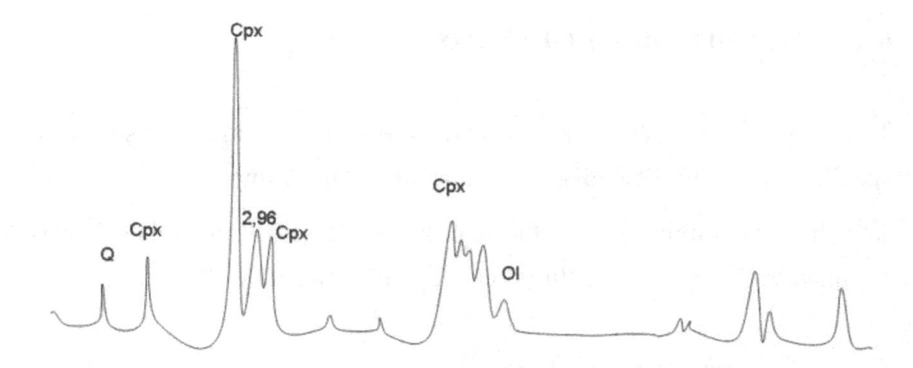

FIGURE 6.1 Radiographs of a specimen subjected to a shock speed of 125 m/s.

X-ray analysis of the siminal sample destroyed at a speed of 650 m / s (Fig. 6.2) revealed that the stishovite peak increased. This is indirect evidence of impact metamorphism. Meanwhile, the intensities of the rest of the peaks also increased, which corresponds to the appearance of impact-induced changes in pyroxene.

Fragments obtained after the destruction of the targets were selected for sample analysis by scanning electron microscopy. Figure 6.3 shows images of the surface fragments of the sample destroyed at a punch speed of 125 m/s.

The surfaces of the sample fragments show two zones: one has a pronounced relief surface, and the other features a smooth, kinked surface (this is typical for amorphous materials). Areas with different morphologies are separated by a small gap, which is likely generated during the process of destruction between the different phases of the cracks. The chemical composition of each area on the surface was determined. It was determined that the area showing "smooth" fracture is composed of SiO_2 (the rest of MgO, FeO); a crystalline fracture in the same area also contains SiO_2 but in much smaller quantities, with albite, magnesite, bauxite, wollastonite, feldspar, and also traces of iron oxide and titanium detected. The crystalline area is heterogeneous; thus, it was studied in greater detail at high magnification, which revealed that the morphology of the surface profile of the first projections of relief coincides with that induced by natural crystal growth. This result confirms that the initial crack was formed directly at the interface between two phases with different degrees of ordering.

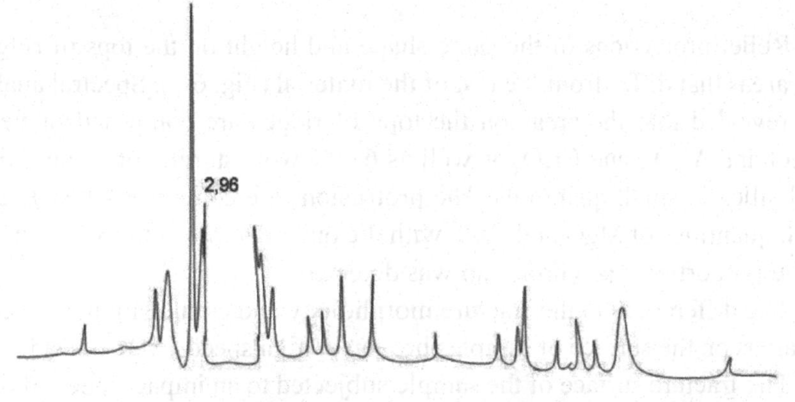

FIGURE 6.2 Spectrogram corresponding to the sample destroyed at an impact speed of 650 m/s.

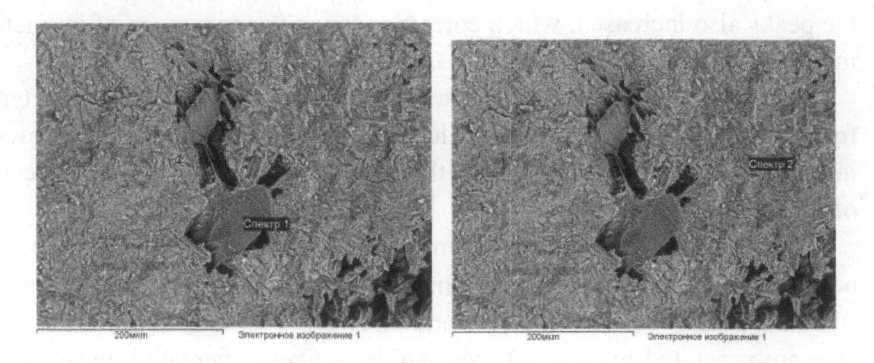

FIGURE 6.3 Surface of fragments of the sample destroyed at a punch speed of 125 m/s.

FIGURE 6.4 Images showing the increase in the surface area of fragments of the sample destroyed at a punch speed of 125 m/s.

Relief projections of the same shape and height on the tops of ridges are areas that differ from the rest of the material (Fig. 6.4). Spectral analysis revealed that the areas on the tops of ridges are composed of FeO, MgO and Al_2O_3, and Cr_2O_3 as well as 6–7% wollastonite, titanium oxide and silica in small quantities. The protrusions are composed of SiO_2 and high quantities of MgO and CaO, with the other components present in the same proportions; no chromium was detected.

The differences in the fracture morphology and composition of selected areas on the surface of samples broken at high speeds were considered.

The fracture surface of the sample subjected to an impact speed of 650 m /s revealed two distinct zones: one, as observed previously, showed a morphology similar to that induced by natural crystal growth, while the other was smooth (Fig. 6.5a). The cracks were more extensive and represented the contours of crystalline aggregates. A clear shift in some areas relative to others was observed. The smooth sections show heterogeneity, and the crystalline zone features relief structures that show smaller differences in height but look similar to the projections.

Spectral analysis revealed that the smooth surfaces are composed of SiO_2 and MgO, as well as wollastonite and the oxides MnO and FeO. A characteristic feature is the presence of manganese oxide, which was not detected in samples obtained after collision at a slower rate. The surface morphology of the zone observed at high magnification (Fig. 6.5b) also revealed clear differences; the slip bands have relief, similar to the fracture surface observed within the same field in the first case, but at high velocities, the formation of wollastonite was observed.

The sample surface showing a crystalline morphology (Fig. 6.5c) is different in composition, with selected chlorine compounds found in small quantities. A more detailed study of the topography of the crystalline area revealed that the height difference was smaller in some areas. The surface is not acutely angled, and the crystallite size is generally smaller than that in the first case. The composition did not change.

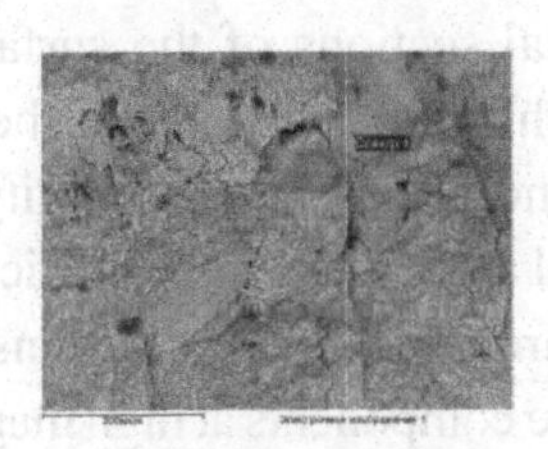

a

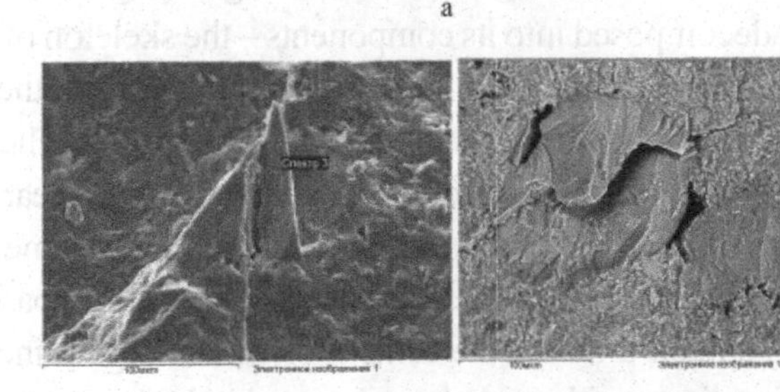

b

FIGURE 6.5 *(Continued)*

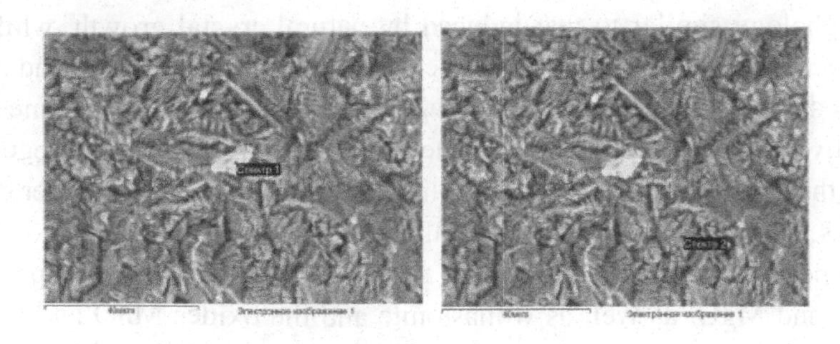

c

FIGURE 6.5 Surface of a fragment of the sample destroyed at a punch speed of 650 m/s: a – general view, b – a zone with slip bands, c – crystalline fracture zone.

The most widespread changes in the structure, surface morphology, and composition of the individual zones were detected in samples obtained by punching at the maximum speed of 800 m/s. First of all, there were no severe fractures; instead, cracking was observed at the folds (all fractures resulted in destruction). Three distinct areas were identified: dark areas of indeterminate morphology (Fig. 6.6), crystalline areas of relief, which were much lower in quantity and smaller, and areas indicating crystallites on the tips of relief structures, many of which were markedly large in size. A study of the individual sections of the surface revealed that the dark areas contain carbon (which was not present before), which is important because carbon forms the basis of the impurity oxides occurring in the material, as well as small amounts of sulfur. Microprobe analysis indicates that these dark patches are most likely thin films.

Studying the rest of the components at high magnification revealed that the shape of the crystallites containing chromium changed (Fig. 6.7). The formed mineral was first decomposed into its components – the skeleton of the crystal. Then, the mineral was deformed and its composition changed; the admixture of vanadium was detected, which was previously undetected. The crystalline mass over much of the surface leveled off at a certain height. Clearly delineating the boundaries of folds surrounding areas of fracture, the same crystallites are clearly distorted and fragmented, despite the fact that their base composition still contains silicon dioxide, the content of which largely increased; the concentration of sodium oxide and chromium oxide increased as a result of the fragmentation of the other phases.

FIGURE 6.6 Surface of a fragment of the sample destroyed at a speed of 800 m/s; allocation of carbon on the fracture surface.

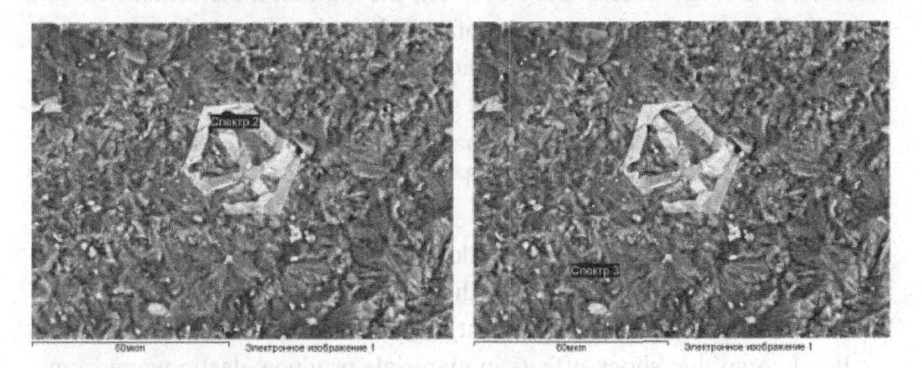

FIGURE 6.7 Destruction and deformation of structural components of siminals with penetration at speeds of 850 m/s.

6.4 DISCUSSION OF THE RESULTS

The results of X-ray diffraction, electron microscopy and microprobe studies suggest that the phenomenon of shock metamorphism occurred in the siminals. However, because the composition of the siminals is a synthetic analog of naturally occurring, shock-induced minerals, the changes consist of those in pyroxene, not in the polymorphic transformations, and the removal and redistribution of elements between the components because pyroxene is resistant to impact.

The pyroxene was composed principally of sodium and calcium, followed by silicon dioxide because of its ability to be transformed into stishovite. When undergoing a polymorphic transition to stishovite, silicon dioxide increases in volume and therefore can be quickly destroyed. It is important to establish why some elements were not detected in the samples before impact, particularly carbon. To understand where it came from, the process through which siminals are obtained should be analyzed.

The raw material used to produce siminals is coal melted by a graphite arc system. The graphite electrodes in the arc system react with the melt such that some carbon goes into the melt, as it it is well dissolved. The interaction between the melt and carbon forms a supersaturated solid solution of carbon in clinopyroxene. A shock load applied to pyroxenes typically removes different elements, in particular, sodium, potassium and calcium, and forges interactions between the siminal material and carbon as well. This explains the presence of vanadium in several phases.

This information allows us to connect all of the changes in the dissipation of mechanical energy in siminal structures in the following sequence:

Shock → polymorphic transformation of quartz into stishovite (swelling of quartz) → local heating and pressure upon the accumulation of clinopyroxene expanding stishovite → Partial disordering of clinopyroxene → allocation of sodium and calcium → allocation of carbon, vanadium, sulfur and other elements of the solid solution and the diffusion of chromium → damage to the fine dust particle melt.

It is known that shock effects in materials produce elastic waves, contributing to the movement of structural defects (such as dislocations). The defects are consolidated; thus, during heat transfer, the vibrations of the surrounding atoms are enhanced. Therefore, during impact, the first phase to suffer the effects is that which exhibits a low degree of long-range order, that is, the amorphous phase. Low-temperature siminals typically feature quartz (a-quartz), which forms trigonal-trapezoidal crystals. The characteristic structural units of silica tetrahedra are connected to each other through their vertices. The tetrahedra form nonplanar six-membered rings in the quartz crystal lattice. In amorphous silica, the tetrahedra are the main structural elements, but there are also eight-membered rings, with the planes of the rings rotated at different angles.

In quartz, the main chemical bond is Si-O. They are approximately equal amounts of homopolar and heteropolar bonds. Thus, the notion that the quartz crystal lattice is composed of Si and O atoms, or $Si_{4-}O_2$ ions, reflects only the limiting cases of real chemical bonds. This suggests the possibility of the homolytic and heterolytic rupture of Si-O upon impact. Quartz in all of its states is a fragile material; thus, upon impact it usually collapses. From mechanosynthesis, it is known that the destruction of quartz due to mechanical shock is similar to its dissolution or melting. According to the phase diagram of quartz, under a sharp blow, the probability of the conversion of quartz to stishovite is enhanced with the decreasing degree of crystallinity. Stishovite has a lower density than the low-temperature version of quartz. Upon impact, the vibrations of elastic waves resonate with the lattice vibrations of quartz. Due to this instantaneous pressure, quartz "swells," which induces stresses arising from relaxation by the formation of cracks. However, not all of the energy goes to the formation of cracks, which leads to heating.

When silicon oxide undergoes a polymorphic transformation, associated the pressure and temperature induce the formation of clinopyroxene in microvolumes. A sufficiently high-pressure force on crystallites of clinopyroxene will lead to intense amorphization. When Na and Ca2 are amorphised, they are pushed out of the matrix, and the effect of temperature will lead to the early melting of microparticles, initiating the decomposition of aluminosilicate tetrahedra.

This process describes the effects of impact on the chromium phase distributed in clinopyroxene. In this process, clinopyroxene is deformed and partially decomposed. Increasing the intensity of exposure leads to transformations in clinopyroxene (which forms a solid solution with carbon). As a result, pure carbon is allocated to the periphery of the crystallites.

6.5 CONCLUSIONS

The results allow us to assert that the effects of impact on siminals have some promise.

First, the effects of impact help remove alkaline impurities. By further remelting powders after impact, better and cleaner siminals can be obtained.

Second, by submitting them to high levels of impact, particles of chromite can be separated for further use as a high-hardness material.

Third, the allocation of carbon in the form of a separate phase suggests that large changes in velocity and pressure conditions can be induced to produce different polymorphs of carbon by the shock metamorphism of siminals.

Thus, the studied changes in the structure of siminals under shock impact form a consistent picture of the materials' behavior. Theoretically, impact could be used as a method of processing siminals to produce unique products.

KEYWORDS

- **chemical composition**
- **high-speed penetration**
- **mineral alloys under impact**
- **rock materials**
- **shock impact**
- **structural changes**

REFERENCES

1. www.unb.ca/passc/ImpactDatabase Website describing 174 meteorite impacts worldwide. Developed and maintained by Planetary and Space Science Centre, University of New Brunswick, Fredericton, New Brunswick, Canada.
2. Collins, G. S., Melosh, J. H., Marcus, R. A., 2005, Earth impact effects program: A web-based computer program for calculating the regional environmental consequences of a meteoroid impact on Earth; Meteorite and Planetary Science 40: 817–840. (www.lpl.arizona.edu/impacteffects)

3. Addison, W. D., Brumpton, G. R., Vallini, D. A., McNaughton, N. J., Davis, D. W., Kissin, S. A., Fralick, P. W., and Hammond, A. L., 2005, Discovery of distal ejecta from the 1850 Ma Sudbury impact event: Geology 33: 193–196.
4. Litasov, K. D., Ohtani, E. (2005) Phase relations in hydrous MORB at 18–26 GPa: Implications for heterogeneity of the lower mantle. Phys Earth Planet Inter 150: 239–263.
5. French, B. M., Koeberl, C., 2010. The convincing identification of terrestrial meteorite impact structures: What works, what doesn't, and why. Earth-Science Reviews, 98, 123–170.
6. Glikson, A. Y., Mory, A. J., Iasky, R. P., Pirajno. F., Golding, S. D., Uysal, I. T. 2005. Australian Journal of Earth Sciences, 52, 545–553.
7. Kumar, R, Kumar, S., BadJena, Mehrotra, S., S. P., Hydration of mechanically activated granulated blast furnace slag, Metallurgical and Materials Transactions B, 36B, 473–484 (2005b).
8. Southworth, D. R., R. Barton, A., S. Verbridge, S., Ilic, B., A. Fefferman, D., H. Craighead, G., J. Parpia, M., "Stress and Silicon Nitride: A Crack in the Universal Dissipation of Glasses, " Physical Review Letters, Vol. 102, p. 4, Jun 2009.
9. Ignatova, A. M., Artemov, A. O., Chudinov V. V., Ignatov, M. N., Sokovnikov, M. A. Issledovanie dissipativnoj sposobnosti sinteticheskih mineral'nyh splavov. IV mezhdunarodnaja konferencija "Deformacija i razrushenie materialov i nanomaterialov." Moskva. 25–28 oktjabrja 2011 g./ Sbornik materialov. M: IMET RAN, 201. 662–664. (http: //dfmn.imetran.ru/)

CHAPTER 7

INVESTIGATION OF EFFICIENCY OF THE INTUMESCENT FIRE AND HEAT RETARDANT COATINGS BASED ON PERCHLOROVINYL RESIN FOR FIBERGLASS PLASTICS

V. F. KABLOV, N. A. KEIBAL, S. N. BONDARENKO,
M. S. LOBANOVA, and A. N. GARASHCHENKO

CONTENTS

7.1 Introduction 214
7.2 Experimental 215
7.3 Results and Discussion 216
7.4 Conclusion 222
Keywords 223
References 223

7.1 INTRODUCTION

The purpose of research was obtaining fire retardant coatings based on perchlorovinyl resin with improved adhesive properties to protect fiberglass plastics. The article presents the results of studies on the influence of a modifier based on the phosphorus-boron-nitrogen-containing oligomer (PEDA) and filler, which is thermal expanded graphite on physical, mechanical and fire retardant properties of the coatings. It was found that the product PEDA is an effective fire retardant, and its introduction to the compositions provides fire resistance and high adhesion of the coatings. The dependence of the filler amount on fire retardant properties of the coatings, the ability to coking and coke strength were identified.

In many cases polymer construction materials are a good alternative to metals and reinforced concrete. Still, it is known that the majority of such materials are combustible. That is why implementation of the materials to the building industry is associated with solving a range of engineering problems; one of them is providing the materials with the required fire safety. The fire hazard of polymers and composite materials is understood as a complex of properties, which along with combustibility includes the ability for ignition, lighting, flame spreading, quantitative evaluation of smoke generation ability, and toxicity of combustion products.

Fiberglass plastics find ever-widening applications in different industrial fields. The main benefit of fiberglass plastics is higher strength and lower density compared to metals; they are not subjected to corrosion. However, together with the valuable property complex of fiberglass plastics, they also have a significant drawback that is a low resistance to open flame.

The sufficient increase in fire safety of fiberglass constructions may be achieved by using the passive protection measures – applying flame retardant intumescent coatings.

Under the influence of high temperatures on a surface of the object protected from fire an intumescent surface appears which obstructs penetration of heat and fire spread over the surface of the material.

For effective fire protection it is necessary to apply compounds which components inhibit combustion comprehensively: in a solid phase it is carried out by transforming the destruction process in a material, in a gaseous phase – preventing the oxidation of the degradation products [1, 2].

A standard formulation of a fire retardant coating includes an oligomer binder as also fire retarding nitrogen, phosphorus and/or halogen containing inorganic and organic compounds. The fire retarding effect is enhanced at the combination of different heteroatoms in an antipyrene [3, 4].

Previously, it was found that phosphorus boron containing compounds are effective antipyrenes in a fire retardant composition [5–7].

In the investigation we developed a new phosphorus-boron-nitrogen-containing oligomer (PEDA). The oligomer has a good compatibility to a polymer binder, slightly migrates from a polymer material, and is an effective antipyrene when has the lower phosphorus content [6, 7].

Phosphorus-boron-nitrogen-containing oligomer and polymer products comprising –P=O, –P–O–B–, –B–O–C– and –C–N–H– bonds are not studied enough. IR spectroscopy showed that these groups are a part of the PEDA macromolecule composition.

To improve physical and mechanical properties of coatings and characteristics of fire protection efficiency, the fire retardant coatings including the phosphorus-boron-nitrogen-containing oligomer PEDA as a modifier and based on perchlorovinyl resin (PVC resin) were obtained. The coatings are used for fiberglass plastics.

7.2 EXPERIMENTAL

With a purpose of defining the efficiency of the developed fire retardant coatings for fiberglass, a set of experiments was conducted.

The experiments on fire retardant properties were carried out on the developed technique by exposure of a coated fiberglass plastic sample to open flame. Time-temperature transformations on the non-heated surface of the fiberglass test sample was registered using a pyrometer measuring the moment of achieving the limit state – a temperature of the fiberglass destruction beginning (280–300°C).

Then, the intumescence index of the coating was calculated. The intumescence index was determined by a relative increase in height of the porous coke layer compared to the initial coating height.

The coke residue was estimated by the relative decrease in the sample weight after keeping it in the electric muffle during 10, 20 and 30 min at 600°C.

For a possibility to use the fire retardant coatings, we need to solve a problem concerned with providing the required adhesion between the coating and protected material. The adhesive strength of the coatings to fiberglass plastic defined as the shear strength of a joint.

During the work, the studies on the combustibility and water absorption of the coatings were conducted.

The combustibility was evaluated by exposure of a sample to the burner flame (temperature peak 840°C) and fixing the burning and smoldering time after fire source elimination.

The experiments on the water absorption were performed in distilled water at temperature 23±2°C for 24 hours. The water absorption was estimated by a sample weight change before and after exposure to water.

The coke microstructure formed after a test sample burning was studied as well.

7.3 RESULTS AND DISCUSSION

As part of the research, the investigation of the coatings based on perchlorovinyl resin and containing the developed intumescent additive PEDA on fire protective properties was conducted. The results are presented in Table 7.1.

TABLE 7.1 Influence of PEDA content on fire resistance of the coatings based on PVC resin.

Parameter	Without coating	PEDA content, wt.%							
		0		2.5		5.0		7.5	
Coating thickness, mm		0.7	1.0	0.7	1.0	0.7	1.0	0.7	1.0
Intumescence index	–	1.55	2.7	4.89	5.55	5.12	6.0	5.64	6.47
Time to the limit state, sec	18	29	32	44	52	48	57	55	63
Temperature of the non-heated sample side in 25 sec, °C	–	247	223	131	115	116	108	109	102

When a coating of 1 mm in thickness containing 7.5% PEDA (% of the initial composition weight) is used, the peak time to the limit state is established, and the intumescence index reaches 6.47.

The temperature dependence of the non-heated sample side on flame exposure time at different PEDA content is shown in Fig. 7.1.

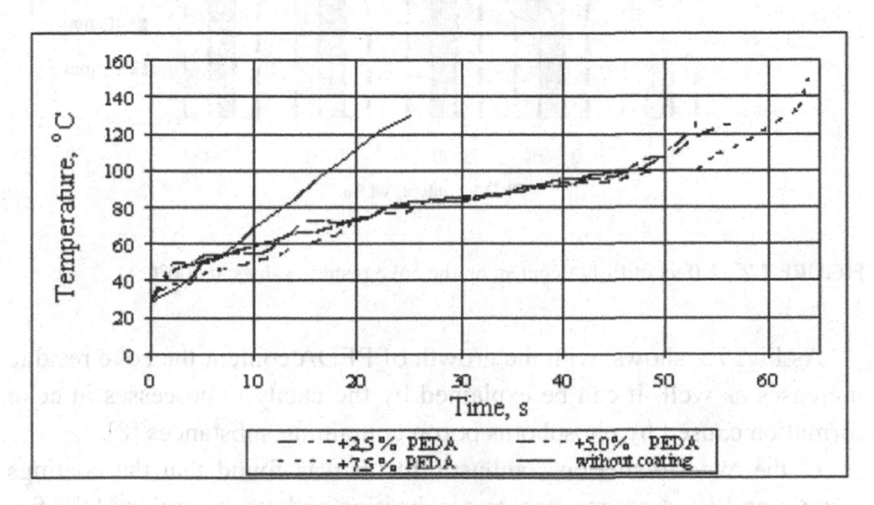

FIGURE 7.1 Dependence of temperature on the non-heated sample side on flame exposure time.

As it is seen in Fig. 7.1, the studied coatings allow keeping temperature on the non-heated sample side within the range 80–100°C for quite a long time; time to the limit state of test samples increased by 2–2.5 times.

Coke formation is an important process in fire and heat protection of a material. Achieving a higher intumescence ratio for carbonized mass, lower heat conductivity of coke and its sufficient strength, all these characteristics are the necessary conditions for effective fire protection.

The dependence of PEDA content influence on ability to form coke is presented on Fig. 7.2.

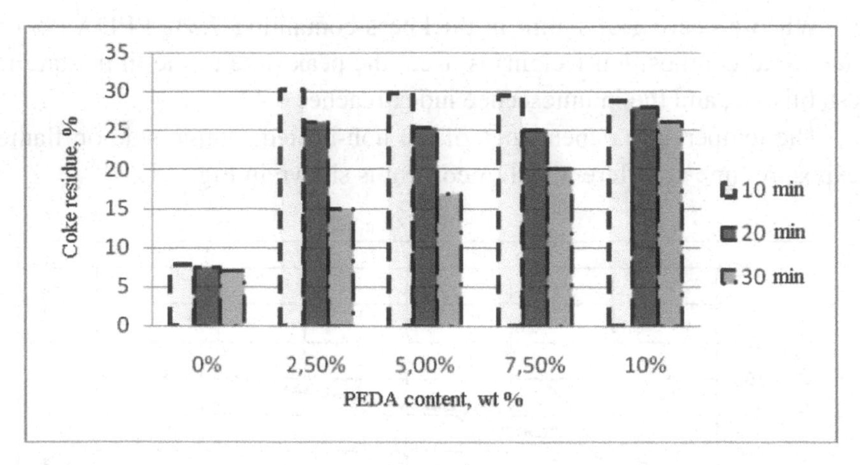

FIGURE 7.2 Effect of PEDA content on the coke residue values at 600°C.

As Fig. 7.2 shows, with the growth of PEDA content the coke residue increases as well. It can be explained by the catalytic processes in coke formation caused by phosphorus boron containing substances [8].

In the experiments on combustibility it was found that the coatings containing PEDA are resistant to combustion and can be assigned the fire reaction class 1 as nonflammable (see Table 7.2).

TABLE 7.2 Influence of PEDA content on combustibility of the coatings based on PVC resin.

PEDA content, wt.%	Combustibility of a coating
0	Burning
2.5	Self-extinguishing in 2 seconds
5.0	Self-extinguishing in a second
7.5	Not burning

The combustibility tests demonstrate that introducing PEDA into the compositions based on PVC resin promotes formation of a large coke layer; the coating film does not burn, because the presence of nitrogen in the modifying additive enables an enhancement of the fire and heat resisting effect.

TABLE 7.3 Influence of PEDA content on water absorption of the coatings based on PVC resin.

PEDA content, wt.%	Extent of change in sample weight	pH
0	0.02	7
2.5	–0.05	5
5.0	–0.06	5
7.5	–0.05	5
10.0	–0.07	4

The results on determining the water absorption of the modified samples revealed an insignificant washout of PEDA that takes place through the slight diffusion of the modifying additive to the film surface that is evidenced by a change in pH in 24 hours (Table 7.3). Nevertheless, this has no effect on fire resistance of the coatings.

As noted above, intumescent coatings should have good adhesion to the protected material; therefore, the studies on the influence of PEDA content on the adhesion strength of the coatings based on PVC resin to the fiberglass plastics were carried out while researching. The test results are illustrated by Fig. 7.3.

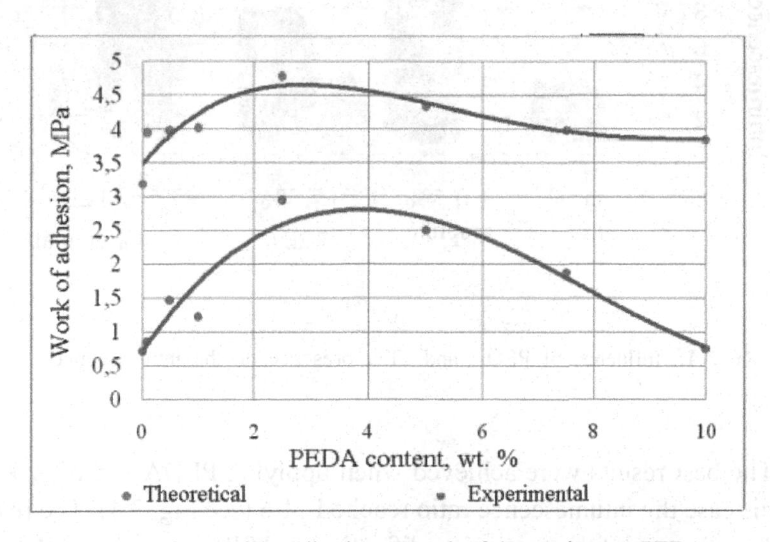

FIGURE 7.3 Dependence of the adhesion strength of the coatings on PEDA content.

So, it was established that the introduction of PEDA to the coating composition in amounts of 2.5–7.5% provides an increase in the adhesion strength by 1.5–4 times.

For confirmation of the experimental data, the work of adhesion was calculated according to the Young-Dupre equation. The surface tension was observed on the duNouy tensiometer. The contact angles of wetting were defined using a goniometric method.

The calculated values of the work of adhesion are well correlated with experimental data.

In order to improve intumescence and fire protection, the effect of introduced thermal expanded graphite (TEG), which served as a filler, on coke formation and physical and mechanical characteristics of the coatings was also studied in the work.

In a course of the study an optimal graphite amount was chosen so that the adhesion characteristics of the coatings would not become worse and allow obtaining a sufficiently hard coke.

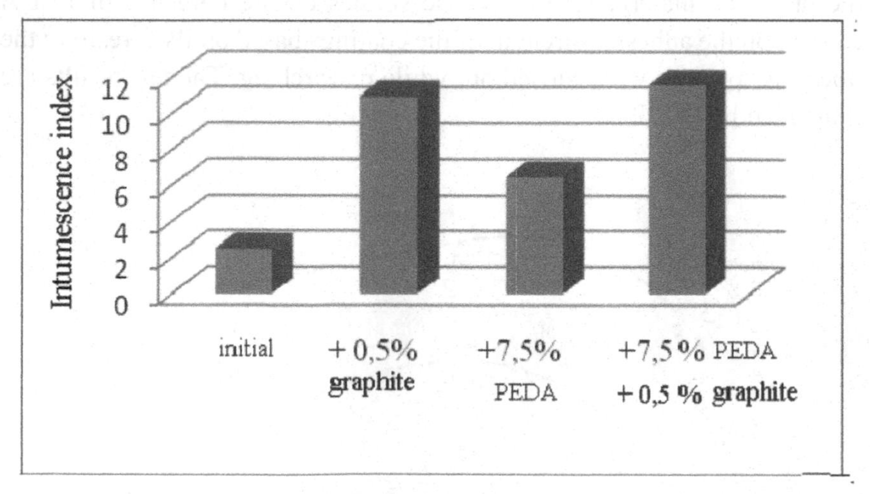

FIGURE 7.4 Influence of PEDA and TEG presence on the intumescence index of coatings.

The best results were achieved when applying PEDA and filler TEG. In this case the intumescence ratio reached 11.6 (see Fig. 7.4). The results on the influence of the coating modification and filler presence on the coke structure are presented in Figs. 7.5–7.7.

The foamed coke formed at the composition testing and not containing modifying additives and fillers has a coarse aorphous structure (Fig. 7.5); there are foamed globular formations of 10–100 μm in the coke volume grouping to associates.

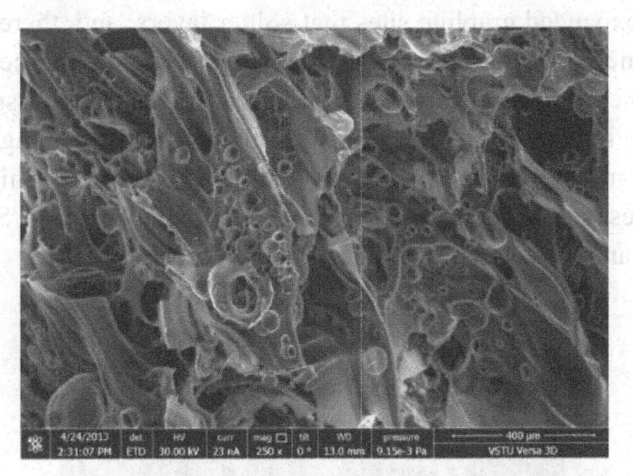

FIGURE 5 Micrograph of a coke structure on the initial coating at 250-fold magnification.

In the compositions containing TEG only (Fig. 7.6), the coke structure is mainly determined by graphite that is present in the form of extended structures longer than 1000 μm, 50–100 times greater than the pore size of the foamed phase. The presence of these structures leads to increased friability of the foamed mass, coke has low strength.

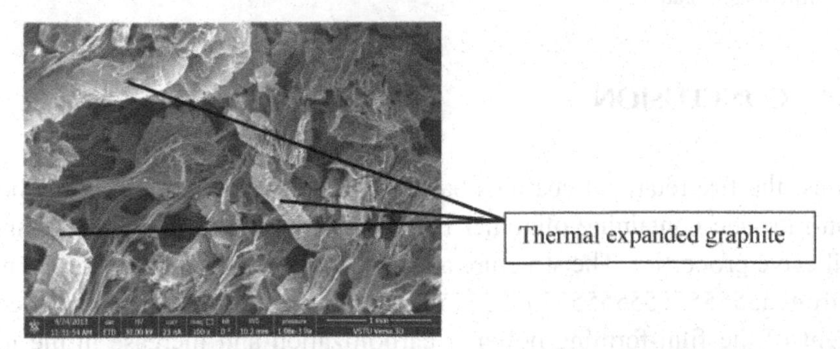

FIGURE 6 Micrograph of a coke structure on the coating containing TEG at 100-fold magnification.

In the coke structure of the coating containing PEDA and TEG (see Fig. 7.7) the extended structures, formed by graphite, disappear, and there are only short fragments of these formations. A consolidation of the carbon layers is observed which, probably, takes place due to formation of the high temperatures of polyphosphoric acids on the surface and between the layers of expanded graphite sites that solder layers, and, thereby, impede TEG intumescence; there is a slight shrinkage of the intumescent layer. As a result, the intumescence index of this composition is not substantially exceed the intumescence index of the composition containing only filler, but with a more ordered structure of coke and a sufficiently high strength and hardness of the composition the fire resistance increases. Such a coke can withstand more intense combustion gas streams.

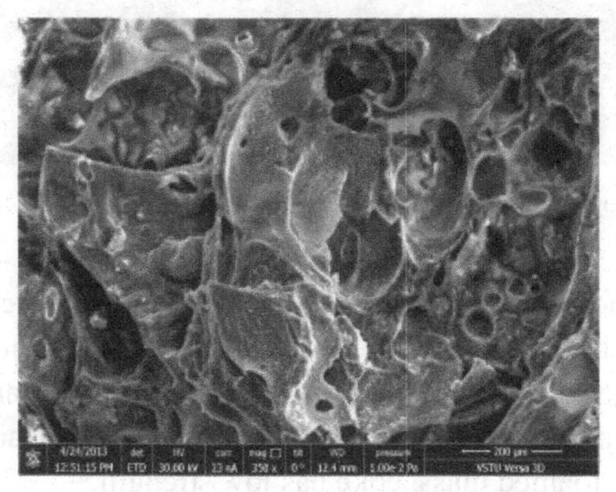

FIGURE 7 Micrograph of a coke structure on the coating containing PEDA and TEG at 350-fold magnification.

7.4 CONCLUSION

Thus, the fire retardant coatings based on the developed phosphorus-boron-nitrogen-containing oligomer have high fire and heat protective and adhesive properties. The structure and presence of phosphorus, boron and nitrogen555555555555555555555555 heteroatoms promote an enhancement of the film-forming polymer carbonization and increase in the intumescence ration of the coatings. In addition, the definite advantage of

PEDA application is that it is slightly washed out of a coating when exposed to water.

Introduction of the modifying additive PEDA in combination with a filler – thermal expanded graphite – permits to increase the coating intumescence by 11 times, resulting in improved fire and heat protective properties of the coatings and reduced destruction of fiberglass plastic.

The research has been done with financial support from Ministry of Education and Science of the Russian Federation under realization of the federal special-purpose program "Scientific, academic and teaching staff of innovative Russia" for 2009–2013 years: The Grant Agreement №14. B37.21.0837 "Development of active adhesive compositions based on element organic polymers and vinyl monomers."

KEYWORDS

- **adhesion**
- **fiberglass plastics**
- **filler**
- **fire protection**
- **fire retardant coatings**
- **modifier**
- **phosphorus-boron-nitrogen-containing oligomer**

REFERENCES

1. Al. Al. Berlin. Combustion of Polymers and Polymer Materials of Reduced Combustibility.Soros Educational Journal. 1996. № 9. 57–63.
2. *Shuklin, S. G., V. Kodolov, I., E. N. Klimenko. Intumescent Coatings and the Processes that Take Place in Them.Fiber Chemistry.* 2004. Vol. 36. Issue 3. 200–205.
3. **Nenakhov,** *S.* **A., V. P. Pimenova. Physical Transformations in** Fire Retardant Intumescent Coatings Based on Organic and Inorganic Compounds.Pozharovzryvobezopasnost Fire and Explosion Safety**. 2011. Vol. 20. № 8. 17–24.**

4. Balakin, V. M., E. Yu. Polishchuk. Nitrogen and Phosphorus Containing Antipyrenes for Wood and Wood Composite Materials.Pozharovzryvobezopasnost Fire and Explosion Safety. **2008. Vol. 17,** № 2. 43–51.
5. Ya, I., Shipovskiy, S. Bondarenko, N., I. Yu. Goryainov. Fire Protective Modification of Wood / Proceedings of the International Scientific and Practical Conference. Dnepropetrovsk, 2005.– Vol. 47 p. 20.
6. Gonoshilov, D. G., N. Keibal, A., S. Bondarenko, N., V. F. Kablov. Phosphorus Boron Containing Fire Retardant Compounds for Polyamide Fibers.Proceedings of the 16th International Scientific and Practical Conference "Rubber industry: Raw materials. Manufactured materials. Technologies," Moscow, 2010 160–162.
7. Lobanova, M.S., V. Kablov, F., N. Keibal, A., S. N. Bondarenko. Development of Active Adhesive Fire and Heat Retardant Coatings for Fiber-Glass Plastics.All the Materials. Encyclopedic Reference Book. 2013. № 04 55–58.
8. Korobeynichev, O. P., A. Shmakov, G., V. M. Shvartsberg. The Combustion Chemistry of Organophosphorus Compounds.Uspekhi khimii. 2007. Vol. 76. № 11. 1094–1121.

CHAPTER 8

MECHANICAL PERFORMANCE EVALUATION OF NANOCOMPOSITE MODIFIED ASPHALTS

M. ARABANI and V. SHAKERI

CONTENTS

8.1 Introduction..226
8.2 Literature Review 227
8.3 Experimental Setup and Procedure...228
8.4 Results and Discussions...232
8.5 Experimental Creep Compliance Models 239
8.6 Conclusions..241
Keywords 242
References...242

8.1 INTRODUCTION

In recent years, many researches were done to improve service life of asphalt pavement against vehicles dynamic loads. For this purpose, researchers investigated different ways such as changing the aggregate gradation and using of additive material to modify bitumen and asphalt mixtures. One of this ways is using of additive materials to improve asphalt properties against dynamic loads. Due to unique characteristics of nano materials, using of them in asphalt mixtures has been interested. Therefore, in this study the effect of nano zinc oxide (ZnO) in improvement of mechanical properties of hot mix asphalt (HMA) has been investigated. To achieve this goal, mixtures with different content of bitumen and nano zinc oxide were made and the effects of these parameters were investigated on the modified mixtures in comparison to conventional asphalt mixtures. With the experimental results and the numerical analysis with Matlab Software, two experimental models were proposed for prediction of the creep behavior of both conventional and modified asphalt mixtures by nano ZnO for different conditions depending on temperature and stress. The results showed that adding of nano zinc oxide to HMA had a great effect in the improvement of permanent deformation of HMA.

Damages that occur before the useful life of pavement mainly are rutting, permanent deformation and fatigue cracking. Since the recovery and reconstruction of defects will be costly, therefore, the prevention of such cases would be more economical. To avoid failure, one of the methods is to modify the properties of bitumen. Researchers have used different methods including the use of various types of polymers and fibers [1].

To improve the performance of bitumen and asphalt concrete mixtures, the addition of modifiers such as nano materials has become popular in recent years. Nano composites are one of the most popular materials discovered to improve properties of bitumen and asphalt mixture [2]. Nanotechnology is the creation of new materials, devices, and systems at the molecular level as phenomena associated with atomic and molecular interactions strongly influence macroscopic material properties [3]. Typically, the Purpose of nanoscale is from about 1 nm to 100 nm. This technology has attracted the attention of many researchers.

Many researches are done on modified bitumen with Nano materials, but it is the first time that modification with nano Zinc oxide (ZnO). Zinc oxide nanoparticles have bigger aspect ratio and a large surface area in comparison with normal zinc oxide, and are not uniform in size and arrangement. It is much more active in terms of chemical. By adding these materials to bitumen, because the bonds that are formed between the material and the bitumen, Properties of bitumen, including the softening point, penetration grade, and ductility are improved. It is expected that modification of bitumen with Nano materials improve the mechanical properties of asphalt mixtures including increase of stiffness modulus, increase of strength against stripping, increase of strength against moisture damage, Prevention of cracks and increase of resistance against creep.

The goal of this study is to evaluate the influence of nano ZnO on the engineering properties of bitumen and asphalt concrete mixtures. For this purpose, they were performed penetration grade, softening point, ductility, and rotational viscometer (RV) tests on modified bitumen by four different content of nano ZnO and repeated load axial (RLA) test on asphalt concrete mixtures by three different content of nano ZnO. With the experimental results and the numerical analysis with Matlab Software, two experimental models were proposed for prediction of the creep behavior of both conventional and modified asphalt mixtures with optimum nano ZnO for different conditions depending on temperature and stress.

The scientific contributions of this paper are:

- Using nano ZnO as modifier of bitumen in hot mix asphalt increases the efficiency of asphalt mixtures.
- It is found that replacing the optimum content of nano ZnO improves the creep behavior of the asphalt mixtures.
- The addition of nano ZnO as modifier of bitumen can improve the creep behavior of asphalt mixtures even at high temperatures and stresses.

8.2 LITERATURE REVIEW

In recent years, many researches were done to improve service life of asphalt pavement against vehicles dynamic loads by using of nano materials. Ghaffarpour et al. [4] carried out comparative rheological tests on bitumen and mechanical tests on asphalt mixtures containing unmodified and nanoclay modified bitumen. Results showed that nanoclay could improve properties of asphalt mixtures such as stability, resilient modulus, and indirect tensile strength, but it do not seem to have a useful effect on fatigue behavior in low temperature. Golestani et al. [5] evaluated Performance of bitumen modified with nano composite. The physical, mechanical and rheological properties of original bitumen, and bitumen modified with nano composite have been studied and compared. The results showed that nano composite could improve the physical properties, rheological behaviors and the stability of the bitumen. Vandeven et al. investigated nanotechnology effects on the adhesion of asphalt mixtures. Two different types of nanoclay were used to modify bitumen. In the first case, viscosity of modified bitumen in comparison to original bitumen (70–100) did not change after the addition of 6% (by weight) of nanoclay, although it was improved its short-term and long-term hardening. In the second case, viscosity of bitumen was increased after adding Nanoclay [6]. Ghasemi et al. [7] evaluated the potential benefits of nano-SiO_2 powder and SBS for the asphalt mixtures used in pavements. Five bitumen formulations were prepared by using various percentages of SBS and nano-SiO_2 powder. Then, Marshall Samples were prepared by the modified and unmodified bitumens. The results of this investigation indicated that the asphalt mixtures modified by 5% SBS plus 2% nano SiO_2 powder could give the best results in the tests. Khodadadi et al. [8] investigated the effect of adding Nanoclay on long-term performance of asphalt mixtures. Indirect tensile test was conducted on cylindrical specimens made of conventional and modified bitumen at the stress levels of 200, 300, 400 and 500 KPa. The results showed that the addition of 2% nanoclay could increase the fatigue life of the asphalt mixtures.

8.3 EXPERIMENTAL SETUP AND PROCEDURE

8.3.1 MATERIALS

The aggregates used in this study were graded using the continuous type IV scale of the AASHTO standard, which is presented in Table 8.1 [9]. Bitumen was a 60–70 penetration grade and its properties are shown in Table 8.2. Also, properties of nano ZnO are shown in Table 8.3 [10].

TABLE 8.1 Gradation of aggregates used in the present study.

Sieve(mm)	19	12.5	4.75	2.36	0.3	0.075
Lower–upper limits	100	90–100	44–74	28–58	5–21	2–10
Passing (%)	100	95	59	43	13	6

TABLE 8.2 Properties of bitumens used in this study.

Test	Standard	Result	Specification limit
Penetration (100 g, 5 s, 25 °C), 0.1 mm	ASTM D5-73	64	60-70
Ductility (25 °C, 5 cm/min), cm	ASTM D113-79	102	Min 100
Solubility in trichloroethylene, %	ASTM D2042-76	99.5	Min 99
Softening point, °C	ASTM D36-76	51	49-56
Flash point, °C	ASTM D92-78	250	Min 232
Loss of heating, %	ASTM D1754-78	0.2	Max 0.8

TABLE 8.3 Properties of nano ZnO used in this study.

Specification	Result
Molecular formula	ZnO
Molecular Weight (gr/mol)	81.4
Color	White
Odor	Odorless
Particle size (nm)	50
Apparent Density (lbs/ft^3)	35
gr/cm^3)) Specific Gravity	5.6
Flash point (°C)	1436
Melting Point (°C)	1975
Boiling Point (°C)	2360
Weight loss after burning (%)	Less than 0.6 %
Solubility In Water (30°C)	0.00016 gr/100 mL

8.3.2 LABORATORY TESTS

8.3.2.1 EMPIRICAL RHEOLOGICAL TESTS ON BITUMENS

To determine the optimum content of Nano ZnO for bitumen, penetration grade, softening point, ductility, and rotational viscometer (RV) tests were carried out on conventional and modified bitumen with different nano ZnO content. The nano ZnO content selected were 2%, 5%, 8%, and 11% by weight of bitumen. The modification of bitumen with nano ZnO was performed at nano scale level by thermodynamic driving force. Tests were performed according to the standard test procedures. The penetration grade test is an empirical test, which measures the consistency (hardness) of asphalt at a specified test condition according to ASTM-D5 standard. This test method covers determination of the penetration of semisolid and solid bituminous materials. Penetration test method is usually done on 25°C and at 5 second. For determination the softening point of bitumen in the range from 30 to 157°C, the ASTM-D36 is used. Also ductility of bitumen is determined according ASTM-D113 standard. This test method provides one measure of tensile properties of bituminous materials and may be used to measure ductility for specification requirements. Rotational viscometer test, as described in ASTM-D4402, is used to measure the apparent viscosity of bitumen from 38 to 260°C. The torque on the apparatus-measuring geometry, rotating in a thermostatically controlled sample holder containing a sample of bitumen, is used to measure the relative resistance to rotation. The torque and speed are used to determine the viscosity of the asphalt in Pascal seconds. In this study, rotational viscometer test is done on 135°C.

8.3.2.2 REPEATED LOAD AXIAL TEST (RLA)

The tests generally used to assess the resistance of asphalt concrete mixtures to permanent deformation are the Marshall test, the static creep test, the dynamic creep test, repeated axial load test (RLA), and the wheel-tracking test [11]. In this study, the resistance to permanent deformation of modified asphalt concrete mixtures by different nano ZnO was evaluated by using RLA test. RLA test has been used for a long time to investigate creep behavior of asphalt mixtures, which is because of its simplicity and logic relation with permanent deformation of asphalt mixture. The most important outcome of RLA test is accumulative strain curve facing number of loading cycles, which depends on compound rutting strength. Fig. 8.1 depicts a form of this curve. As shown in Fig. 8.1, the curve was made of three major parts: primary stage with relatively large deformation during a short number of cycles; secondary stage which the rate of accumulation of permanent deformation remains constant; and tertiary stage, the final stage that the rate of deformation accelerates until complete failure takes place. This stage is usually associated with the formation of cracks [4].

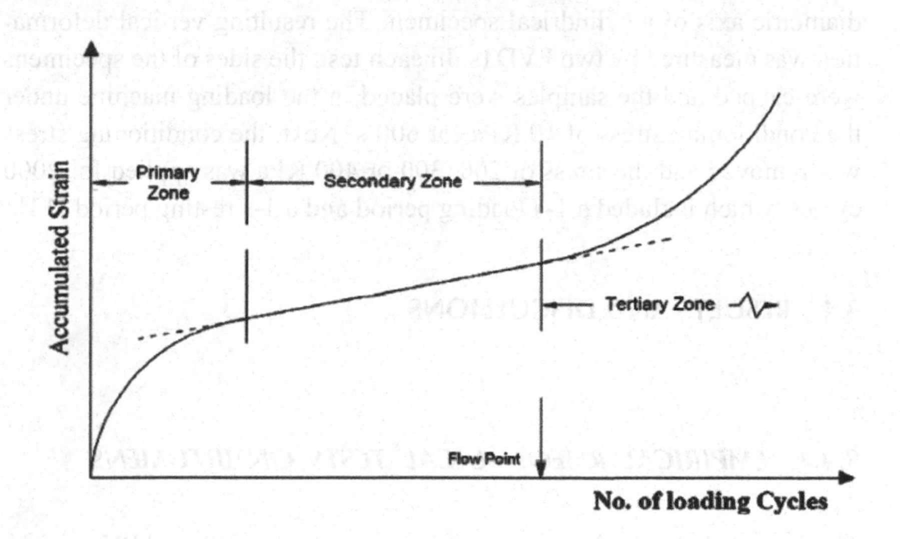

FIGURE 8.1 A typical creep curve.

For this study, cylindrical samples were made with a diameter of 101 mm and the height is 70 mm. Specimen preparation and compaction were conducted in accordance with ASTM D1559 [12]. To determine the optimum content of Nano ZnO for asphalt mixtures, the nano ZnO content selected were 2%, 5% and 8% by weight of bitumen. The program for the prepared samples and the experimental tests is described in Table 8.4.

TABLE 8.4 Program for specimen preparation and testing.

Parameter	Levels
Percentage of Nano ZnO	(0, 2 , 5, 8)%
Test temperatures	(40, 45, 50, 55, 60) °C
Type of gradations	Topka
Stress	200, 300 and 400 (KPa)

The creep compliance was determined by applying a dynamic compressive load (repeated load axial) of fixed magnitude of 200, 300 or 400 KPa for 1h at different temperature 40, 45, 50, 55, and 60°C along the diametric axis of a cylindrical specimen. The resulting vertical deformation was measured by two LVDTs. In each test, the sides of the specimens were capped and the samples were placed in the loading machine under the conditioning stress of 10 KPa for 600 s. Next, the conditioning stress was removed and the stress of 200, 300 or 400 KPa was applied for 2000 cycles, which included a 1-s loading period and a 1-s resting period [13].

8.4 RESULTS AND DISCUSSIONS

8.4.1 EMPIRICAL RHEOLOGICAL TESTS ON BITUMENS

The results of penetration grade, softening point, ductility and RV tests are presented in Figs. 8.2–8.5. It can be seen from figures that adding of nano ZnO has positive effect on the rheological properties of bitumen. The viscosity of bitumen is increased and Penetration of bitumen is decreased by adding the nano ZnO. Moreover, by decreasing temperature sensitivity of modified bitumen due to adding nano ZnO, the softening point of bitumen is improved. The decrease and increase in penetration and softening point demonstrate the increased hardness and stiffness of the modified bitumen [14].

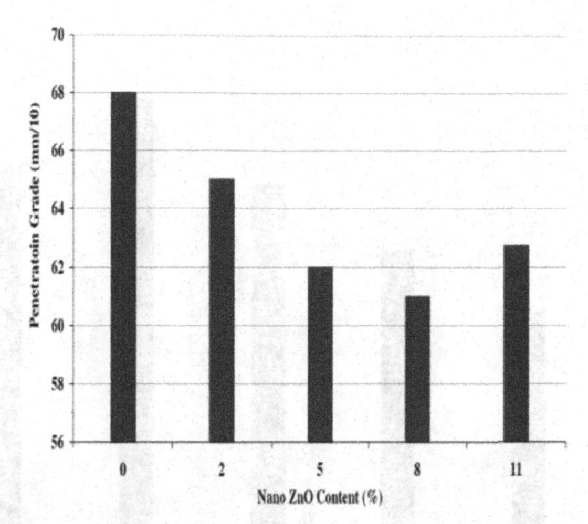

FIGURE 8.2 Penetration grade test results on unmodified and modified bitumens.

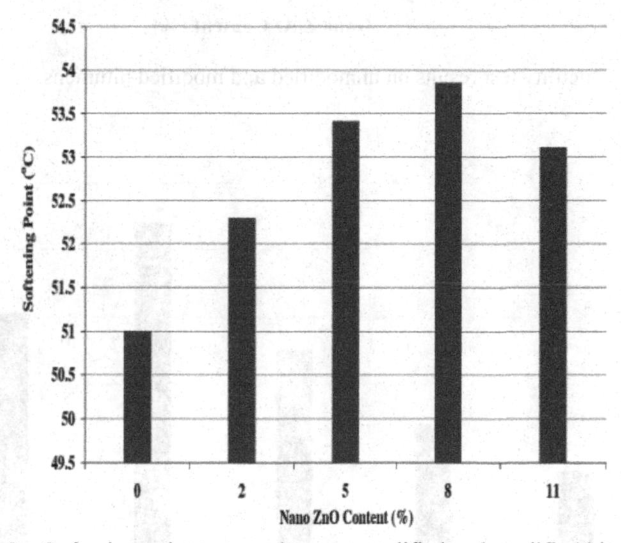

FIGURE 8.3 Softening point test results on unmodified and modified bitumens.

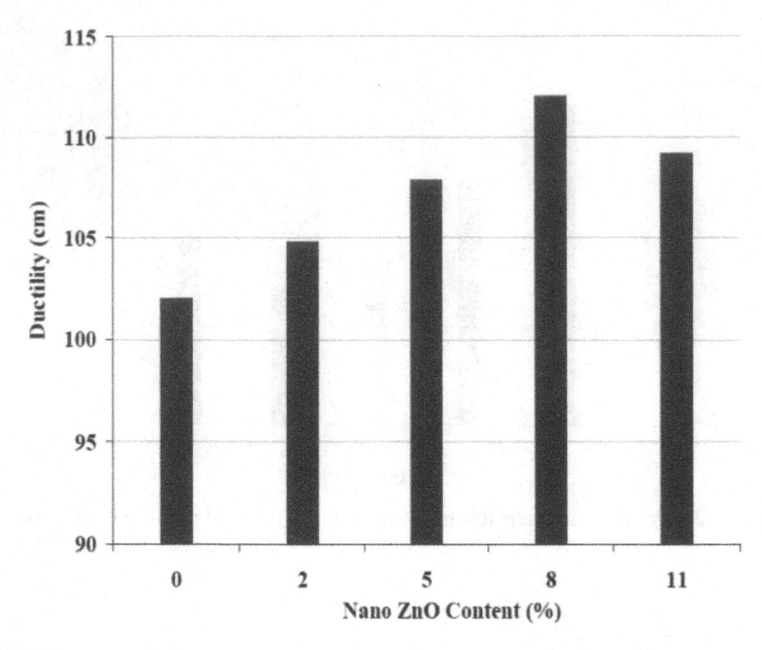

FIGURE 8.4 Ductility test results on unmodified and modified bitumens.

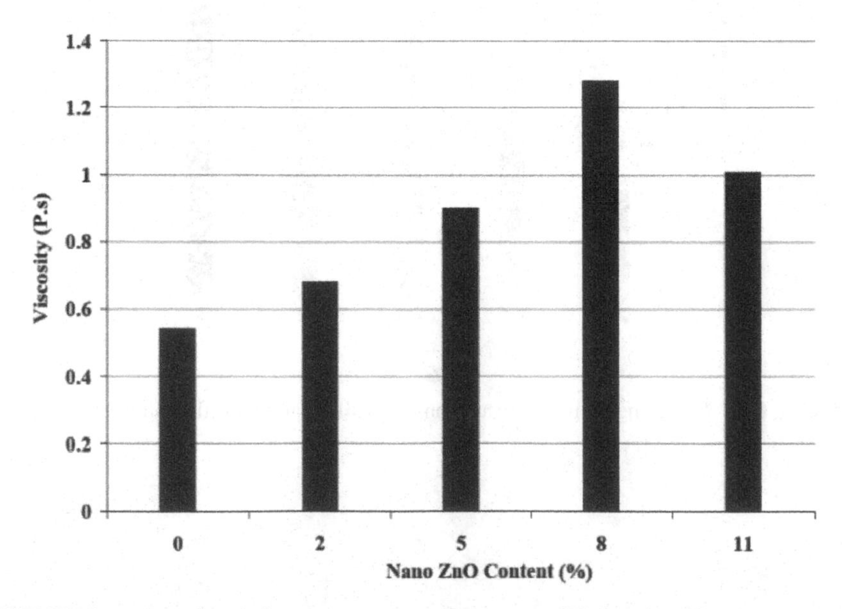

FIGURE 8.5 Rotational viscometer test results on unmodified and modified bitumens.

It is illustrated in Fig. 8.4 and 8.5 that ductility and apparent viscosity (in RV test) are significantly increased to 8% nano ZnO with improvement of modified bitumen stiffness in comparison to conventional bitumen. It can be seen from results by increasing bond between bitumen and nano, viscosity of bitumen is increased to 8% nano content.

The results obtained by the penetration grade, softening point, ductility, and RV tests for bitumen showed that in 11% nano ZnO, penetration grade is increased and ductility, apparent viscosity, and softening point are decreased. As a result, 8% nano ZnO as modifier of bitumen is an optimal content.

8.4.2 REPEATED LOAD AXIAL TEST (RLA)

The values of final strain versus Nano ZnO content in asphalt specimens at different stresses and temperatures are shown in Figs. 8.6–8.10.

The results of the RLA tests show that the samples without nano ZnO have more permanent deformation than the samples containing nano ZnO as modifier of bitumen.

The amount of final strain at a special temperature for samples with different containing nano ZnO is less than conventional samples. According to Figures, it can be concluded that the final strain of mixtures is decreased due to greater adhesion between aggregate and modified bitumens in asphalt mixture in comparison to conventional asphalt mixtures. It can be seen from figures, increasing of temperature is caused significant increasing in the amount of samples final strain.

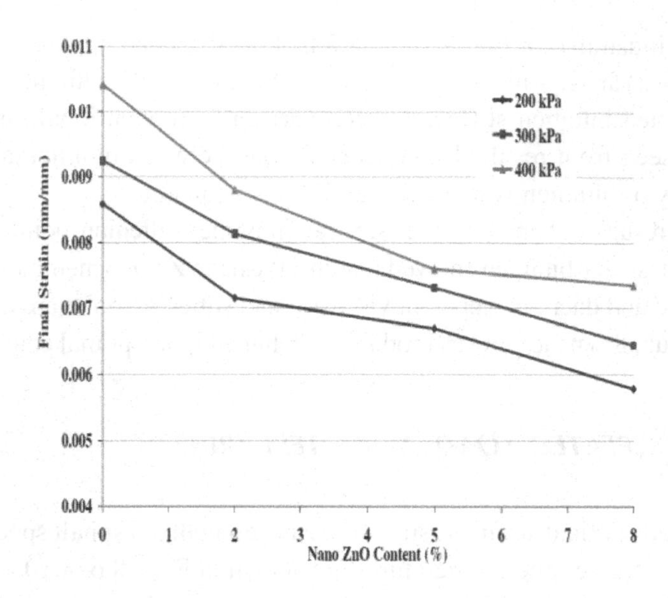

FIGURE 8.6 Variation of final strain versus Nano ZnO content in asphalt specimens at 40°C.

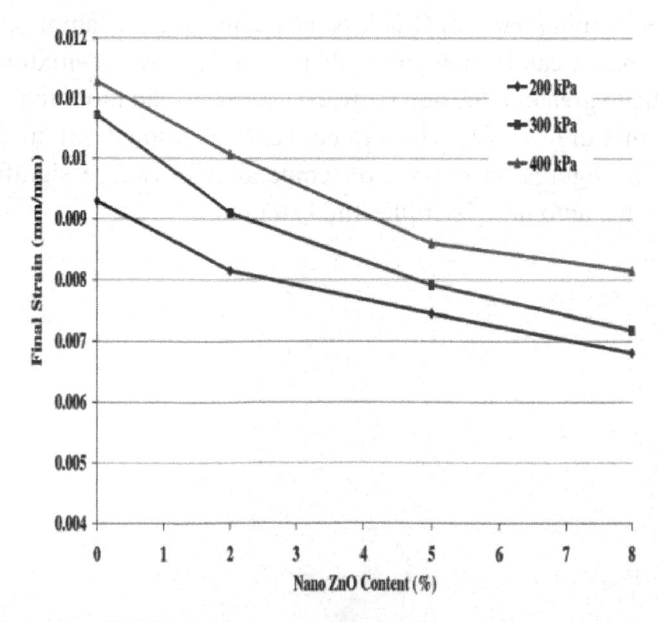

FIGURE 8.7 Variation of final strain versus Nano ZnO content in asphalt specimens at 45°C.

For a special mixture like modified sample with 5% nano ZnO, the amount of final strain at 60°C was about 1.28 times of final strain at 40°C. Also, another remarkable point was that the process of decreasing final strain due to increase of nano content would be less by increasing temperature.

Because of the high sensitivity of the bitumens to the variations of temperature, the final strain and permanent deformation of the conventional and modified mixtures increased at higher temperatures. This phenomenon can be explained by the viscosity and stiffness modulus of the bitumens, which decreased at higher temperatures.

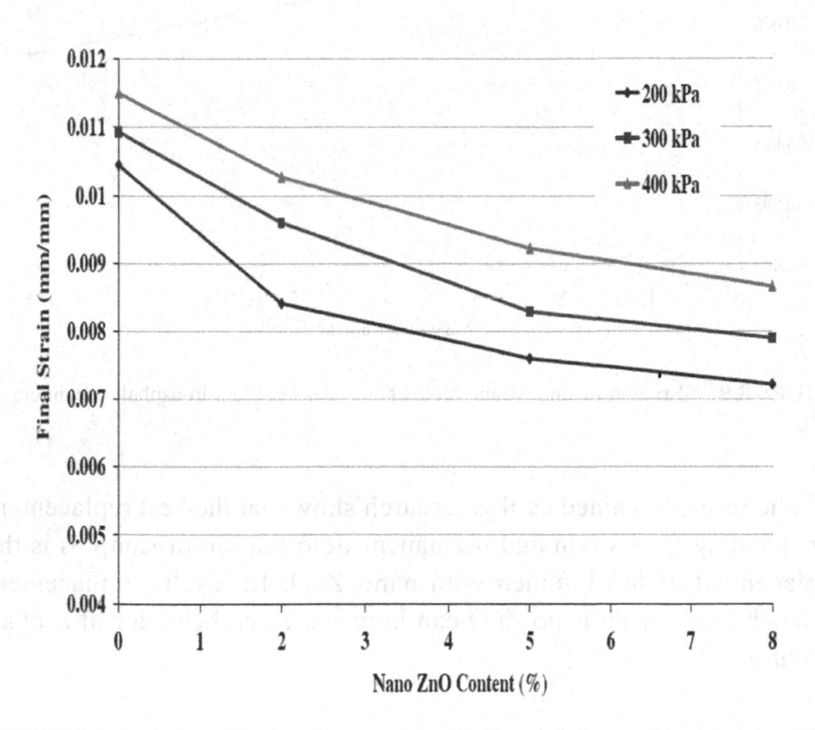

FIGURE 8.8. Variation of final strain versus Nano ZnO content in asphalt specimens at 50°C.

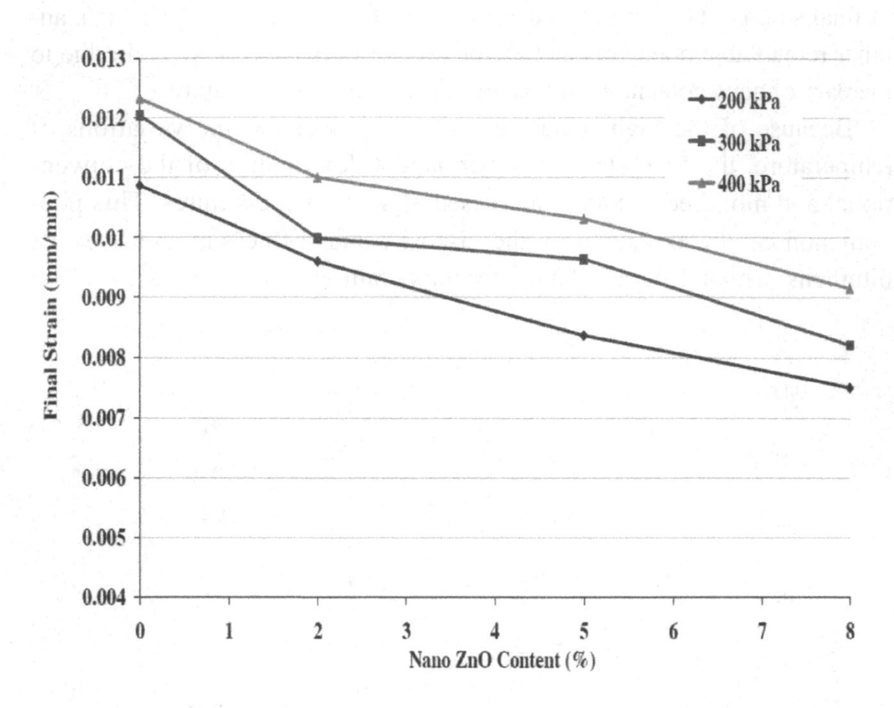

FIGURE 8.9 Variation of final strain versus Nano ZnO content in asphalt specimens at 55°C.

The results obtained by this research show that the best replacement for reducing final strain and permanent deformation of samples is the replacement of 8% bitumen with nano ZnO. In results, replacement of 8% bitumen with nano ZnO can improve creep behavior of asphalt mixtures.

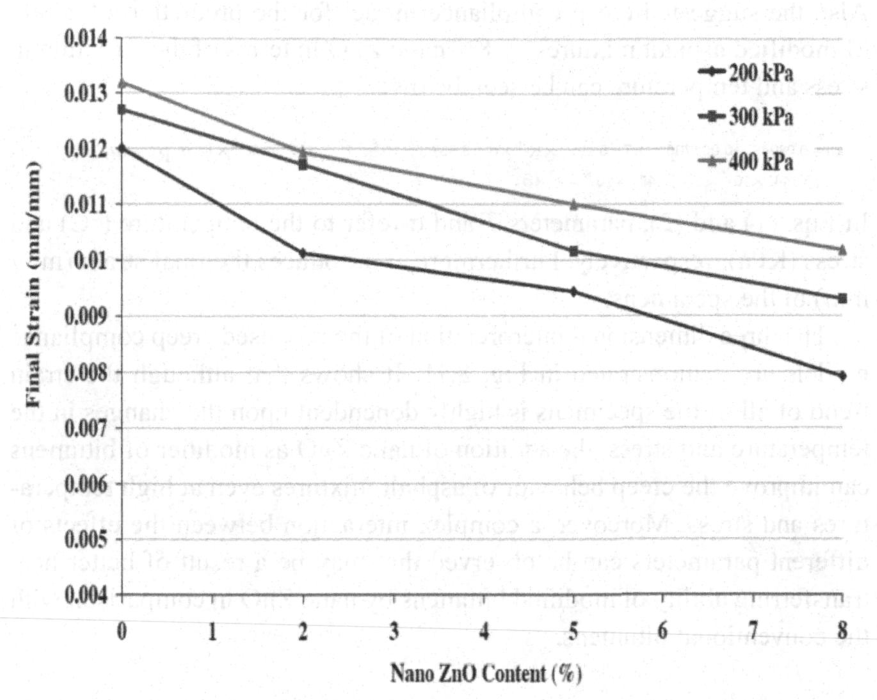

FIGURE 8.10 Variation of final strain versus Nano ZnO content in asphalt specimens at 60°C.

8.5 EXPERIMENTAL CREEP COMPLIANCE MODELS

After determination optimum content of nano ZnO, With the experimental results and the numerical analysis by Matlab Software, two experimental models are proposed for prediction of the creep behavior of both conventional and modified mixtures only for nano ZnO content 8% (optimum) for different conditions depending on temperature and stress.

The suggested creep compliance model for the prediction of the final Strain of conventional asphalt mixtures in terms of the variation in stress and temperature can be seen below:

$$\varepsilon(T,\sigma) = -0.05365 + 0.00329 \times T + 3.391 \times 10^{-5} \times \sigma - 6.009 \times 10^{-5} \times T^2 - 8.901 \times 10^{-7} \times T \times \sigma \quad 3.86 \times 10^{-7} \times T^3 + 7.101 \times 10^{-9} \times T^2 \times \sigma \qquad R^2 = 97.6\% \quad (1)$$

Also the suggested creep compliance model for the prediction of Strain of modified asphalt mixtures by 8% nano ZnO in terms of the variation in stress and temperature can be seen below:

$$\varepsilon(T,\sigma) = -0.0804 + 0.004706 \times T + 6.195 \times 10^{-5} \times \sigma - 8.537 \times 10^{-5} \times T^2 - 2.371 \times 10^{-6} \times T \times \sigma \quad R^2 = 96.4\% \qquad (1)$$
$$5.144 \times 10^{-7} \times T^3 + 2.543 \times 10^{-8} \times T^2 \times \sigma$$

In Eqs. (1) and (2), parameters *T* and σ refer to the temperature (°C) and stress (KPa), respectively. Furthermore, ε introduces the final strain (mm/mm) of the specimens.

The three dimensional interpretation of the proposed creep compliance models are demonstrated in Fig. 8.11. It shows that although the strain trend of all of the specimens is highly dependent upon the changes in the temperature and stress, the addition of nano ZnO as modifier of bitumens can improve the creep behavior of asphalt mixtures even at high temperatures and stress. Moreover, a complex interaction between the effects of different parameters can be observed that may be a result of better heat transferring ability of modified bitumens by nano ZnO in comparison with the conventional bitumens.

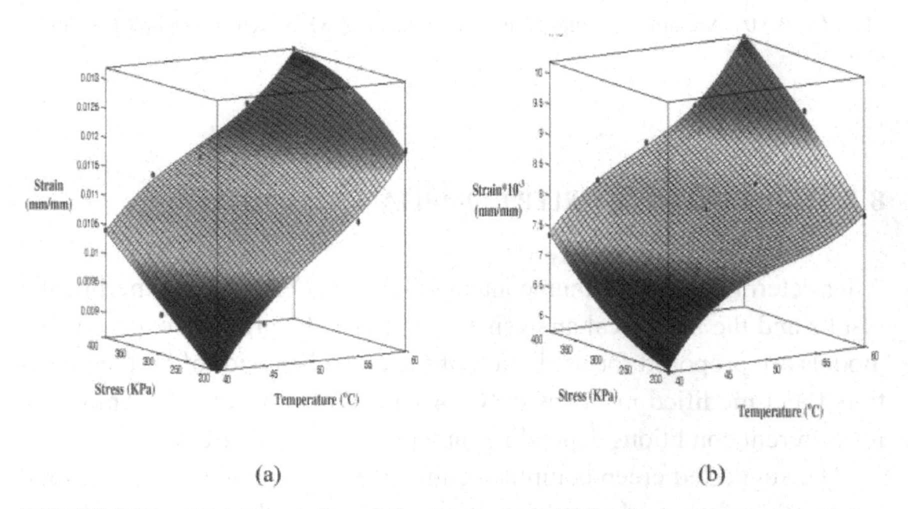

FIGURE 8.11 3D interpretation of the proposed model for final strain vs. temperature and stress in a) conventional asphalt mixture b) Modified asphalt mixture by 8% nano ZnO.

8.6 CONCLUSIONS

In this study, modeling by analytical tools was used for the first time for predicting the creep behavior of modified asphalt mixtures by nano ZnO. Asphalt mixtures with different nano ZnO content and at five different temperatures were subjected to experiments with various levels of stress. Comparison of the proposed models for conventional and modified asphalt mixtures showed that specimens containing nano ZnO have noticeably better creep performance. In addition:

- The results obtained by the penetration grade, softening point, ductility and rotational viscometer tests for bitumen show that 8% nano ZnO as modifier of bitumen is an optimal content.
- The results obtained by the repeated load axial tests for samples show that 8% nano ZnO as modifier of bitumen is an optimal content in asphalt mixtures.
- Using nano ZnO as modifier of bitumen in hot mix asphalt increases the efficiency of asphalt mixtures. It is found that replacing 8% of the bitumen by nano ZnO improves the creep behavior of the asphalt mixtures.
- By increasing the temperature, final strain of all specimens increases. This behavior results from high sensitivity of bitumen in asphalt mixtures to temperature.
- Three dimensional interpretation of the proposed creep compliance models showed that although the strain trend of all of the specimens is highly dependent upon the changes in the temperature and stress, the addition of nano ZnO as modifier of bitumens can improve the creep behavior of asphalt mixtures even at high temperatures and stresses.

KEYWORDS

- **hot mix asphalt (HMA)**
- **macroscopic material properties**
- **mechanical performance**
- **modified asphalts**
- **nano materials**
- **nanocomposite**

REFERENCES

1. Kok, B. V., Kuloglu, N. The Effects of Different Binders on Mechanical Properties of Hot Mix Asphalt. International Journal of Science and Technology 2007; 2(1), 41–48.
2. Pinnavaia, T. J., Beall, G. W., Wiley, J., Sons LTD. Polymer–Clay Nanocomposites. UK, Polymer International 2000; 51(5), 464.
3. Arabani, M., Haghi, A. K., Tanzadeh, R. Laboratory Study on the Effect of Nano-SiO_2 on Improvement Fatigue Performance of Aged Asphalt Pavement. 4th International Conference on Nanostructures (ICNS4), Kish Island, I., R. Iran: 2012.
4. Ghafarpoor, S., Andalibizade, B., Vossough, S. H., Engineering Properties of Nanoclay Modified Asphalt Concrete Mixtures. The Arabian Journal for Science and Engineering 2010; Volume 35, Number 1B.
5. Golestani, B., Moghadas Nejad, F., and Sadeghpour, S. Performance evaluation of linear and nonlinear nanocomposite modified asphalts. Construction and Building Materials 2012; 35: 197–203.
6. Vandeven, M. F., Molenaar, A. A., Besamusca, J., Noordergraaf, J. Nanotechnology for Binders of Asphalt Mixtures. Transportation Research Board 2008; 12(1), 401–412.
7. Ghasemi, M., Marandi, S. M., Tahmooresi, M., Kamali, J., Taherzade, R. Modification of Stone Matrix Asphalt with Nano-SiO_2. Journal of Basic and Applied Scientific Research 2012; 2(2), 1338–1344.
8. Khodadadi, A., and Kokabi, M. Effect of Nanoclay on Long-Term Behavior of asphalt concrete pavement. 2nd Congress of Nanomaterials. University of Kashan. I. R. Iran: 2007.
9. AASHTO guide for design of pavement structures; 1993.
10. Arabani, M. Effect of glass cullet on the improvement of the dynamic behavior of asphalt concrete. Construction and Building Materials 2011; 25: 1181–1185.
11. Niazi, Y., and Jalili, M. Effect of Portland cement and lime additives on properties of cold in-place recycled mixtures with asphalt emulsion. Construction and Building Materials 2009; 23(3), 1338–1343.

12. ASTM D 1559. Standard Test Method for Marshal Test. Annual Book of ASTM Standards, American Society for Testing and Materials, West Conshohocken: 2002.
13. Arabani, M., Moghadasnejad, F., and Azarhoosh, A. R., Laboratory Evaluation of Recycled Waste Concrete into Asphalt Mixtures. International Journal of Pavement Engineering 2012; First article, 1–9.
14. Ghasemi, M., Marandi, S., Tahmooresi, M., kamali, R., Taherzade, R., 2012. Modification of Stone Matrix Asphalt with Nano-SiO2. Journal of Basic and Applied Scientific Research, 2(2), 1338–1344.

CHAPTER 9

MICROSTRUCTURAL COMPLEXITY OF NATURAL AND SYNTHETIC GRAPHITE PARTICLES

HEINRICH BADENHORST

CONTENTS

9.1 Introduction 246
9.2 Materials and Methods 248
9.3 Microstructure of Graphite 248
9.4 Conclusions 277
Keywords 278
References 278

9.1 INTRODUCTION

Graphite is an industrially important raw material with many different applications. However, significant uncertainty still surrounds the microstructure of the multitude of different graphite materials. Through the use of oxidation to expose the underlying microstructure and high-resolution surface imaging it is possible to distinguish between graphite materials from different origins, irrespective of their treatment histories. The key microstructural features can be identified and linked to characteristics such as crystallite sizes and defects. This enables a direct comparison of materials for selection based on a specific purpose. A wide variety of complex microstructures may be found in both natural and synthetic graphite. These are the result of multiple concurrent factors such as processing methods, crystallinity, particle geometry, catalytically active impurities and the accumulation inactive particles. A comprehensive understanding of the intricate microstructural attributes of different graphite materials is critical for the correct interpretation of kinetic and other surface area related parameters. This is by no means an exhaustive study of all possible morphologies found in natural and synthetic graphite materials but it does present an outline of the complex structures that are possible in conjunction with the origin of these structures.

Graphite in its various forms is a very important industrial material, it is used in a wide variety of specialized applications. These include high temperature uses where the oxidative reactivity of graphite is very important, such as electric arc furnaces and nuclear reactors. Graphite intercalation compounds are used in lithium ion batteries or as fire retardant additives. These may also be exfoliated and pressed into foils for a variety of uses including fluid seals and heat management.

Graphite and related carbon materials has been the subject of scientific investigation for longer than a century. Despite this fact there is still a fundamental issue that remains, namely the supra-molecular constitution of the various carbon materials [1, 2]. In particular it is unclear how individual crystallites of varying sizes are arranged and interlinked to form the complex microstructures and defects found in different bulk graphite materials.

Natural graphite flakes are formed under high pressure and temperature conditions during the creation of metamorphosed siliceous or calcareous sediments [3]. Synthetic graphite on the other hand is produced via a multistep, re-impregnation process resulting in very complex microstructures and porosity [4]. For both of these highly graphitic materials the layered structure of the ideal graphite crystal is well established [5]. However, in order to compare materials for a specific application the number of exposed, reactive edge sites are of great importance.

This active surface area (ASA) is critical for quantifying properties like oxidative reactivity and intercalation capacity. The ASA is directly linked to the manner in which crystalline regions within the material are arranged and interconnected. The concept of ASA has been around for a long time [6–11], however due to the nature of these sites and the very low values of the ASA for macrocrystalline graphite, it is difficult to directly measure this parameter accurately and easily. Hence it is been difficult to implement in practice and an alternative method must be employed to assess the microstructures found in graphite materials.

New developments in the field of scanning electron microscopy (SEM) allow very high resolution imaging with excellent surface definition [12, 13]. The use of high-brightness field-emission guns and in-lens detectors allow the use of very low (~ 1kV) acceleration voltages. This limits electron penetration into the sample and significantly enhancing the surface detail, which can be resolved. Due to the beneficiation and processing of the material, graphite exhibits regions of structural imperfection, which conceal the underlying microstructure. Since oxygen only gasifies graphite at exposed edges or defects, these regions may be largely removed by oxidation, leaving behind only the underlying core flake structure. This oxidative treatment will also reveal crystalline defects such as screw dislocations.

Furthermore, the oxidative reactivity of graphite is very sensitive to the presence of very low levels of impurities that are catalytically active. The catalytic activity is directly dependent on the composition of the impurity not just the individual components. Hence a pure metal will behave differently from a metal oxide, carbide or carbonate [14–16], a distinction, which is impossible to ascertain from elemental impurity analysis. The very low levels required to significantly affect oxidative properties are also close to the detection limits of most techniques, as such the only way to irrefutably verify the absence of catalytic activity is through visual inspection of the oxidized microstructure.

Thus, in combination these two techniques are an ideal tool for examining the morphology of graphite materials. Only once a comprehensive study of the microstructure of different materials has been conducted can their ASA related properties be sensibly compared.

9.2 MATERIALS AND METHODS

Four powdered graphite samples will be compared. The first two are proprietary nuclear grade graphite samples, one from a natural source (NNG) and one synthetically produced material (NSG). Both samples were intended for use in the nuclear industry and were subjected to high levels of purification including halogen treatment. The ash contents of these samples were very low, with the carbon content being >99.9 mass %. The exact histories of both materials are not known. The third graphite (RFL) was obtained from a commercial source (Graphit Kropfmühl AG Germany). This is a large flake, natural graphite powder and was purified by the supplier with an acid treatment and a high temperature soda ash burn up to a purity of 99.91 mass %. A fourth sample was produced for comparative purposes by heating the RFL sample to 2700°C for 6 hours in a TTI furnace (Model: 1000–2560–FP20). This sample was designated PRFL since the treatment was expected to further purify the material. All thermal oxidation was conducted in a TA Instruments SDT Q600 thermogravimetric analyzer (TGA) in pure oxygen. The samples were all oxidized to a burn-off of around 30%, at which point the oxidizing atmosphere was rapidly changed to inert. SEM images were obtained using an ultra-high resolution field-emission microscope (Zeiss Ultra Plus 55 FEG-SEM) equipped with an in-lens detection system.

9.3 MICROSTRUCTURE OF GRAPHITE

9.3.1 FEGSEM RESOLUTION

Initially the RFL sample was only purified up to a temperature of 2400°C. When this sample was subsequently oxidized, the purification was found to have been only partially effective. It was possible to detect the effects of trace levels of catalytic impurities, as can be seen from Fig. 9.1.

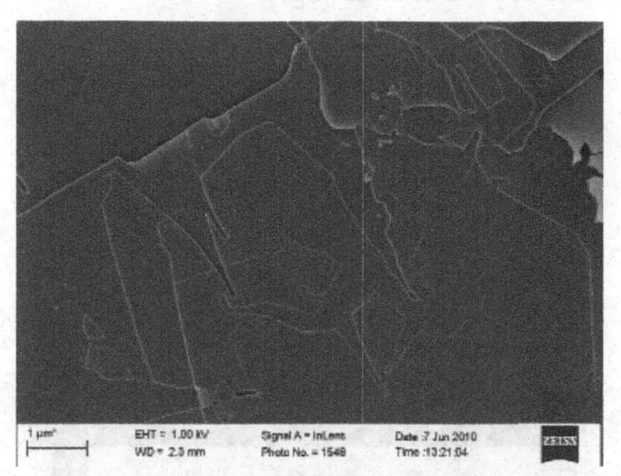

FIGURE 9.1 FEGSEM image of partially purified RFL (30k × magnification).

Since the catalyst particles tend to trace channels into the graphite, as seen in Fig. 9.2, the consequences of their presence can be easily detected. As a result it is possible to detect a single, minute catalyst particle, which is active on a large graphite flake. This effectively results in the ability to detect impurities that are present at extremely low levels.

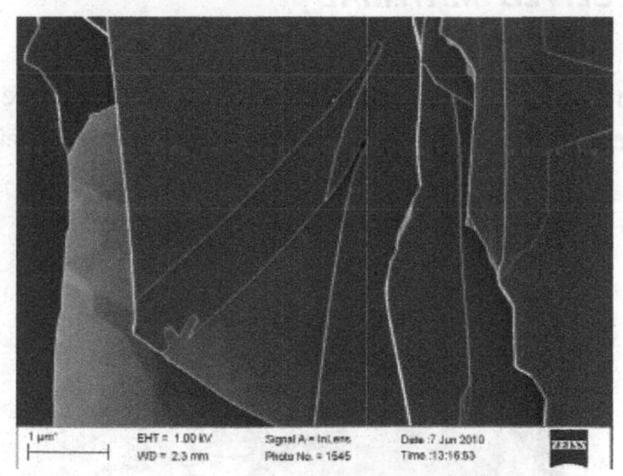

FIGURE 9.2 Channeling catalyst particles (40k × magnification).

When the tips of these channels are examined, the ability of the high resolution FEGSEM, operating at low voltages, to resolve surface detail and the presence of catalytic particles is further substantiated. As can be seen from Fig. 9.3, the microscope is capable of resolving the catalyst particle responsible for the channeling.

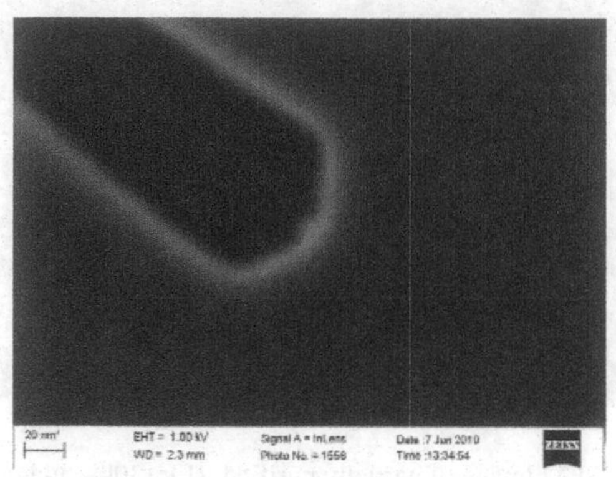

FIGURE 9.3 Individual catalyst particle (1000k × magnification).

In this case the particle in the image has a diameter of around ten nanometers. This demonstrates the powerful capability of the instrument and demonstrates its ability to detect the presence of trace impurities.

9.3.2 AS-RECEIVED MATERIAL

When the as-received natural graphite flakes are examined in Fig. 9.4, their high aspect ratio and flat basal surfaces are immediately evident.

FIGURE 9.4 Natural graphite flakes (175 × magnification).

When some particles are examined more closely, they were found to be highly agglomerated, as shown in Fig. 9.5.

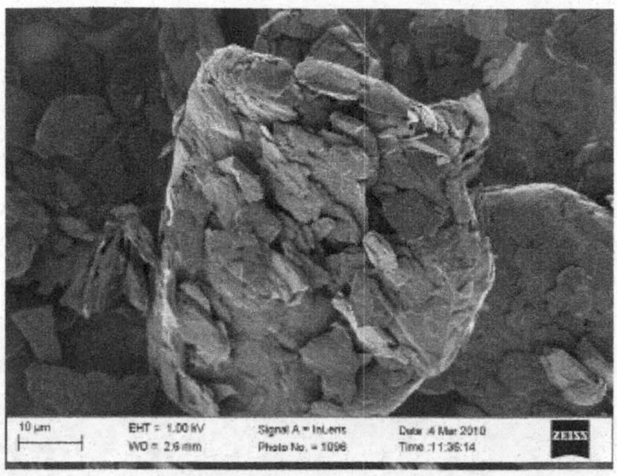

FIGURE 9.5 Close-up agglomerated flake (3k × magnification).

All samples were subsequently wet-sieved in ethanol to break-up the agglomerates. When the sieved flakes are examined they are free of extraneous flakes but still appear to be composite in nature with uniform edges, as can be seen in Fig. 9.6.

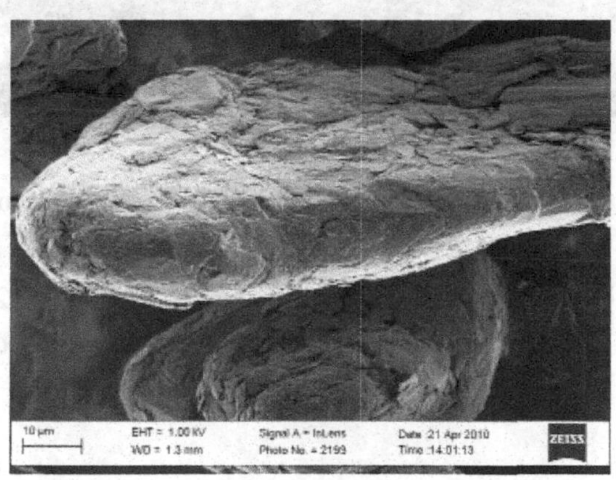

FIGURE 9.6 Close-up of sieved flake (3k × magnification).

This surface deformation is due to the beneficiation process, during which the edges tend to become smooth and rounded. As such the expected layered structure is largely obscured.

9.3.3 PURIFIED NATURAL GRAPHITE

However, when the oxidized natural graphite flakes are examined, their layered character is immediately apparent as seen in Fig. 9.7.

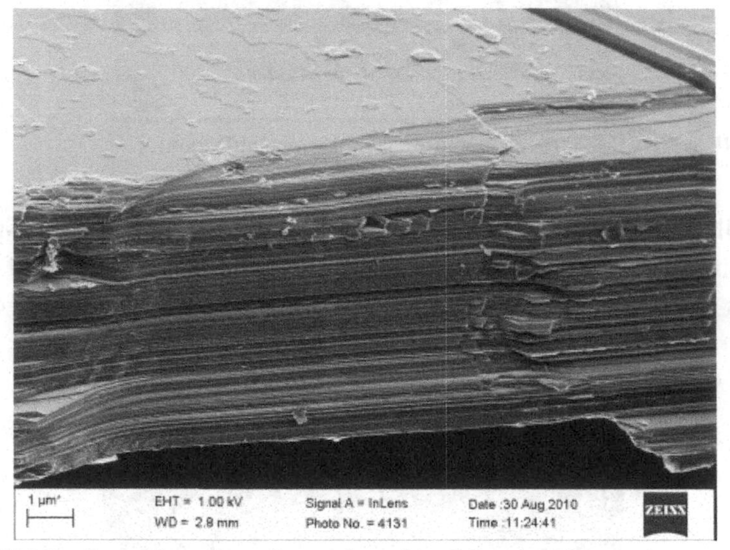

FIGURE 9.7 Layered structure of natural graphite (20k × magnification).

When the edges are examined from above, the crisp 120° angles expected for the hexagonal crystal lattice of graphite are evident as in Fig. 9.8.

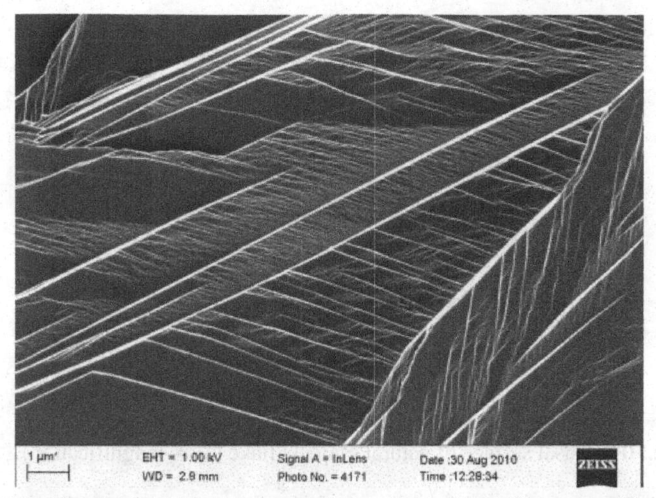

FIGURE 9.8 Hexagonal edge structures of natural graphite (20k × magnification).

The flat, linear morphology expected for a pristine graphite crystal is now more visible in Fig. 9.9.

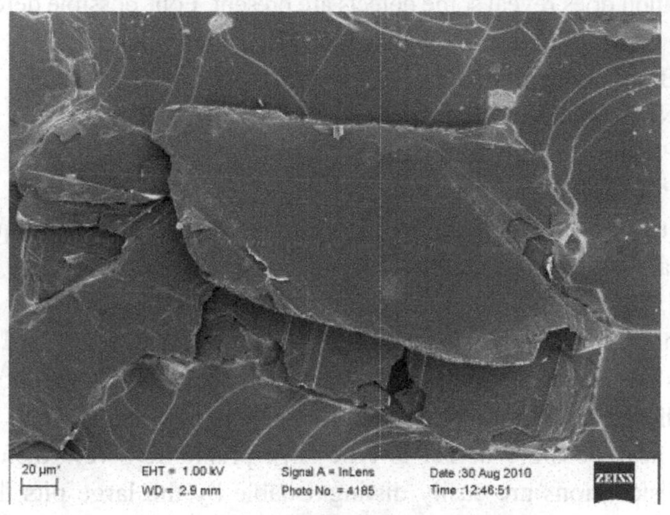

FIGURE 9.9 Oxidized natural graphite flake (800 × magnification).

When the basal surface is examined more closely as in Fig. 9.10, the surface is smooth and flat across several tens of micrometers.

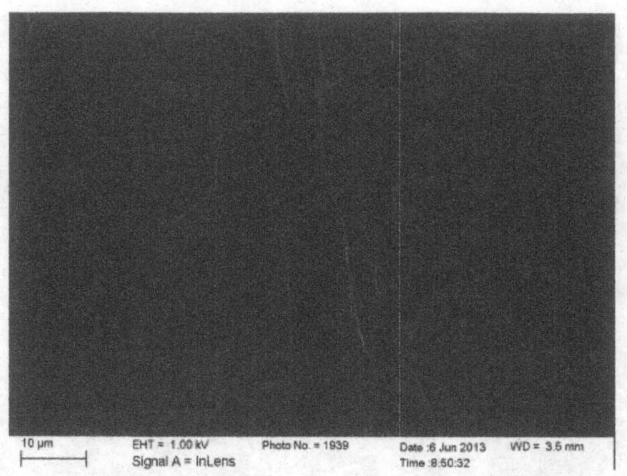

FIGURE 9.10 Basal surface of natural graphite flake (3k × magnification).

Since the graphite atoms are bound in-plane by strong covalent bonding, the basal surface is expected to be comparatively inert. This surface shows no signs of direct oxidative attack, with only some minor surface steps visible. Thus the highly crystalline nature of the material is readily evident but further investigation does reveal some defects are present. Four possible defect structures are generally found in graphite [46, 47], namely:

i) Basal dislocations
ii) Non-basal edge dislocations
iii) Prismatic screw dislocations
iv) Prismatic edge dislocations

Due to the fact that the breaking of carbon-carbon bonds is required for non-basal dislocations, the existence of type (ii) defects is highly unlikely [17, 18]. Given the very weak van der Waals bonding between adjacent layers, however, basal dislocations of type (i) are very likely and a multitude has been documented [5, 18]. However, such defects will not be visible in the oxidized microstructure.

The next possible defect is type (iii), prismatic screw dislocations. These dislocations are easily distinguishable by the large pits that form during oxidation with a characteristic corkscrew shape [19], as visible in Fig. 9.11.

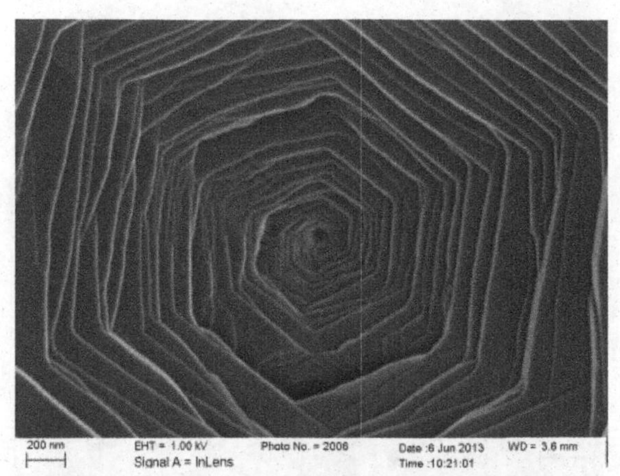

FIGURE 9.11 Prismatic screw dislocation (90k × magnification).

In general, no more than a few discrete occurrences of these defects were found in any given flake. A more prevalent defect is twinning, which is derived by a rotation of the basal plane along the armchair direction of the graphite crystal. These defects usually occur in pairs, forming the characteristic twinning band visible in Fig. 9.12.

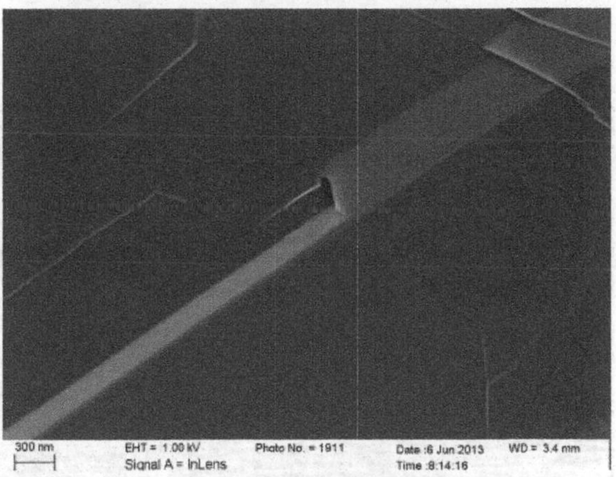

FIGURE 9.12 Twinning band (60k × magnification).

The angled nature of these defects is more evident when they are examined edge-on as in Fig. 9.13.

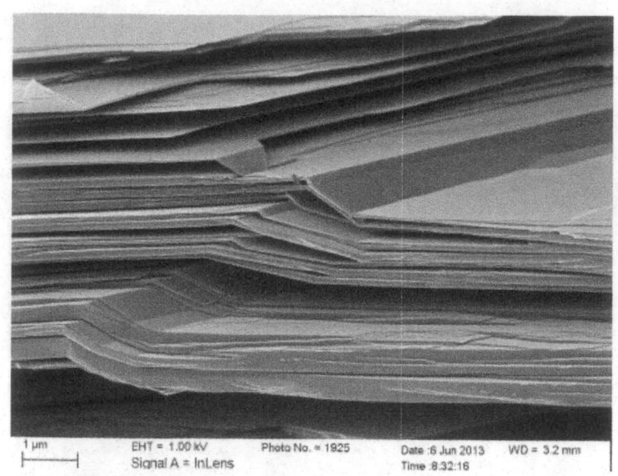

FIGURE 9.13 Twinning band edge (26k × magnification).

These folds are usually caused by deformation but may also be the result of the formation process whereby impurities became trapped within the macro flake structure and were subsequently removed by purification. As is visible in the lower left hand corner of Fig. 9.13, a single sheet can undergo multiple, successive rotations and as can be seen in Fig. 9.14, the rotation angle is variable.

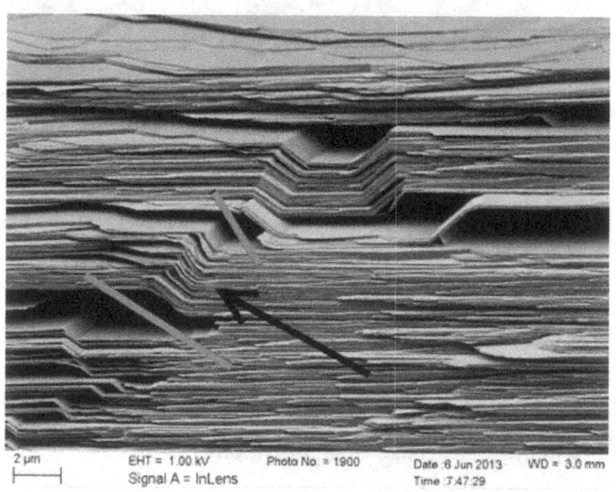

FIGURE 9.14 Rotation of twinning angle (11k × magnification).

The final defect to be considered is prismatic edge dislocations. These involve the presence of an exposed edge within the flake body, an example is shown in Fig. 9.15.

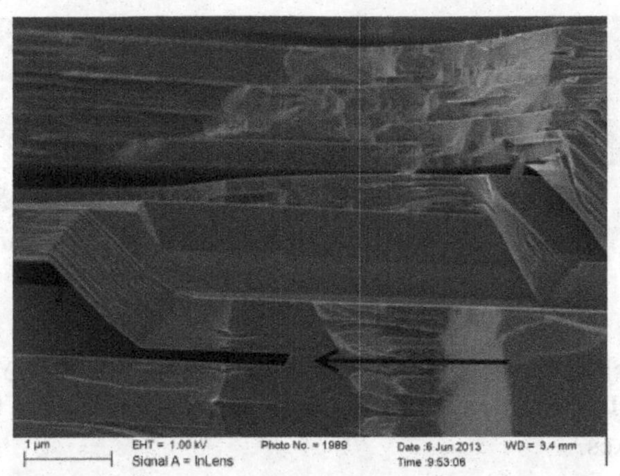

FIGURE 9.15 Prismatic edge dislocation (40k × magnification).

If the edge is small enough the structural order may be progressively restored, leading to the creation of a slit shaped pored that gradually tapers away until it disappears. An example of this behavior is demonstrated in Fig. 9.16.

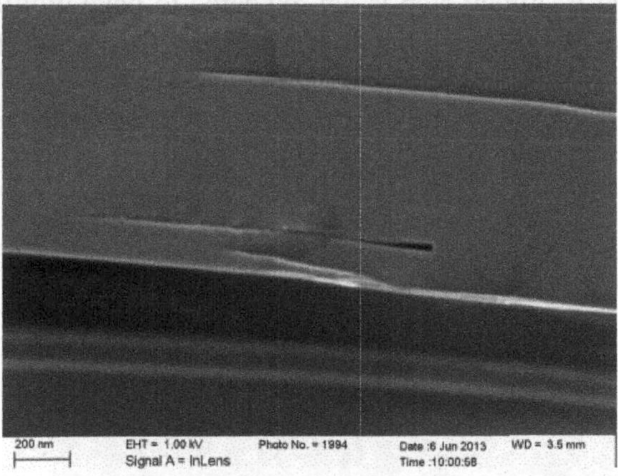

FIGURE 9.16 Gradual disappearance of small edge dislocation (125k × magnification).

If the edge dislocation is larger, when the stack collapses it will lead to a surface step, analogous to a twinning band. This can result in the formation of some very complex structures, such as the one shown in Fig. 9.17.

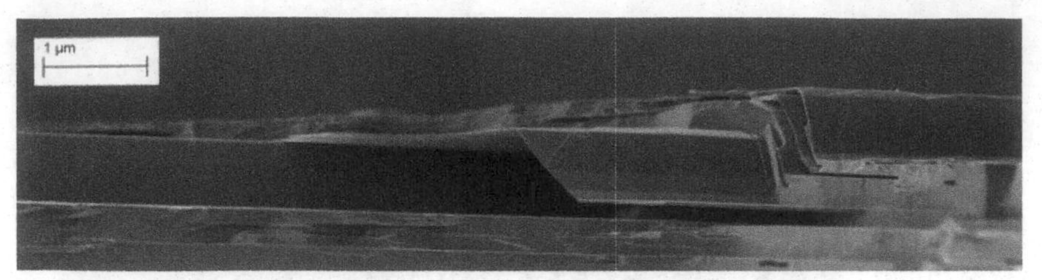

FIGURE 9.17 Complex surface structures (50k × magnification).

Thus despite being highly crystalline with an apparently straightforward geometry, complex microstructures can still be found in these natural graphite flakes.

9.3.4 CONTAMINATED NATURAL GRAPHITE

A very different microstructure is evident when the same natural graphite flakes are examined which have not been purified. Since the flakes are formed under geological processes involving high temperatures and pressures, the heat treatment step is not expected to have modified the flake microstructure. As expected the high aspect ratio and general flat shape of the flakes are still visible in Fig. 9.18.

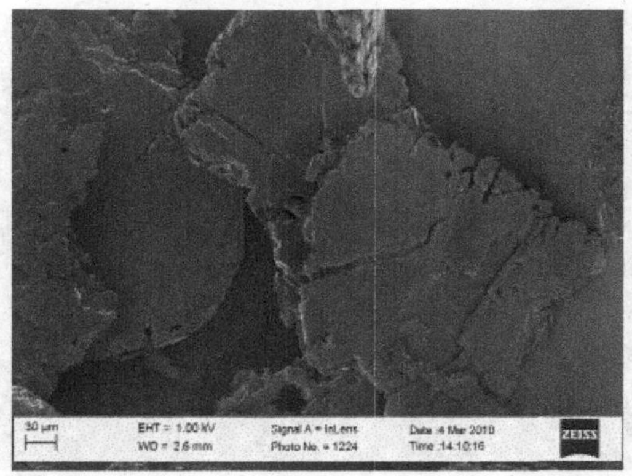

FIGURE 9.18 Flake structure of contaminated natural graphite flakes (500 × magnification).

However, when the edges of these particles are examined more closely as in Fig. 9.19, highly erratic, irregular edge features are observed.

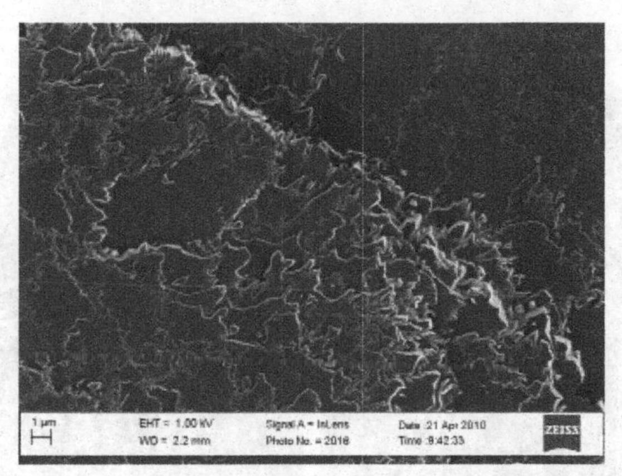

FIGURE 9.19 Erratic edge of contaminated natural graphite flakes (10k × magnification).

When the edges are scrutinized more closely, as in Fig. 9.20, the reason for these edge formations becomes clear. They are caused by minute impurities, which randomly trace channels into the graphite.

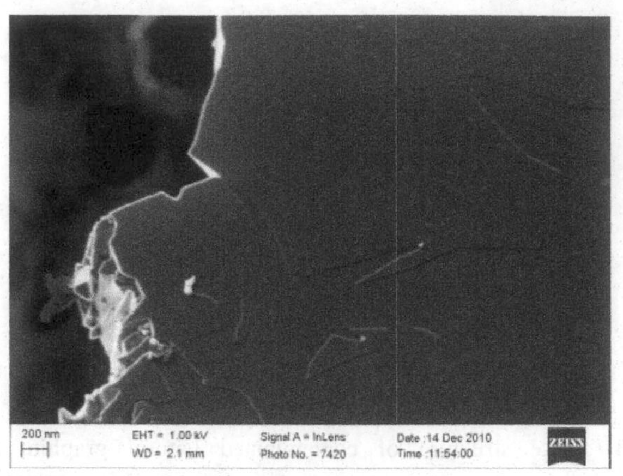

FIGURE 9.20 Catalyst activity (65k × magnification).

In certain cases, the activity is very difficult to detect, requiring the use of excessive contrast before they become noticeable as shown in Fig. 9.21 (A and B).

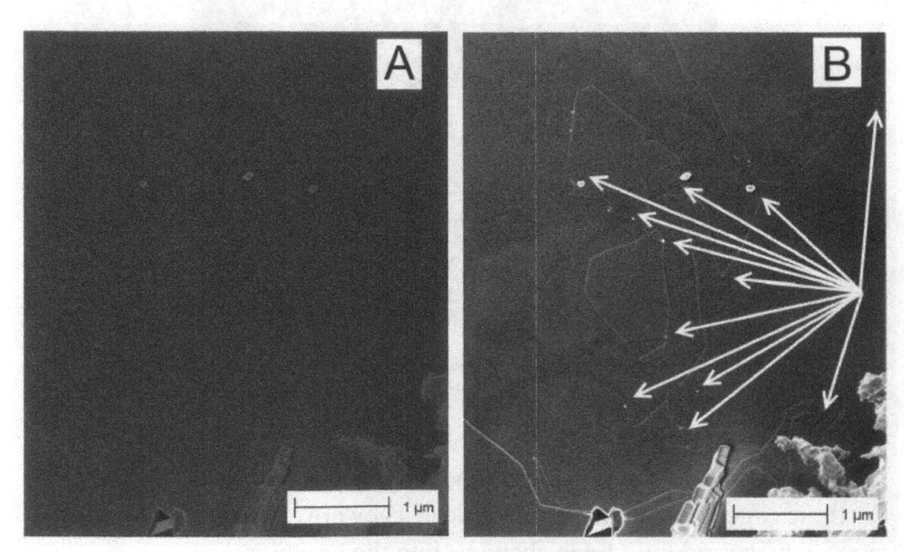

FIGURE 9.21 Contrast detection of catalyst activity (38k × magnification).

A very wide variety of catalytic behaviors were found. Broadly, these could be arranged into three categories. The first, show in Fig. 9.22, are small, roughly spherical catalyst particles. Channels resulting from these particles are in most cases triangular in nature. In general it was found that these particles tend to follow preferred channeling directions, frequently executing turns at precise, repeatable angles, as demonstrated in Fig. 9.22B. However, exceptions to these observed behaviors were also found, as illustrated in Fig. 9.22C.

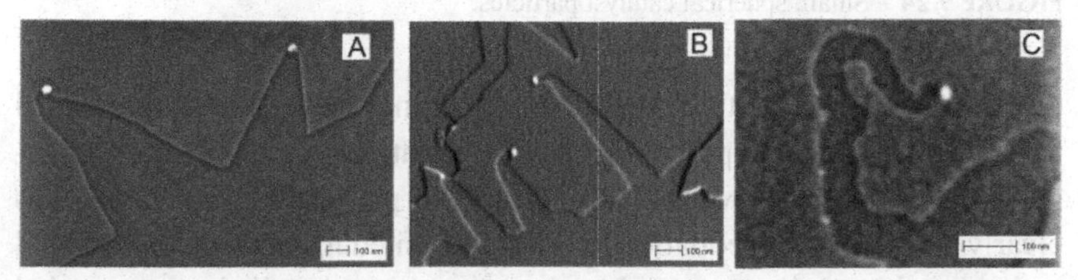

FIGURE 9.22 Small, spherical catalyst particles.

The second group contained larger, erratically shaped particles, some examples of which are shown in Fig. 9.23.

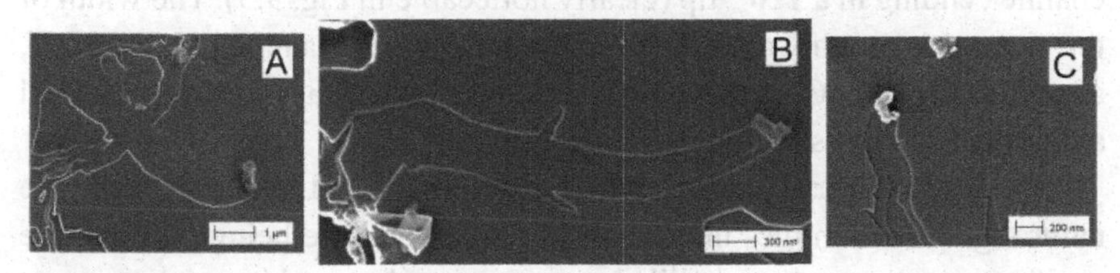

FIGURE 9.23 Small, spherical catalyst particles.

These particles exhibited random, erratic channeling. Where it is likely that the previous group may have been in the liquid phase during oxidation, this is not true for this group, since the particles are clearly capable of catalyzing channels on two distinct levels simultaneously, as can be seen in Fig. 9.23 (B and C). The final group contains behaviors, which could not be easily placed into the previous two categories, of which examples are shown in Fig. 9.24.

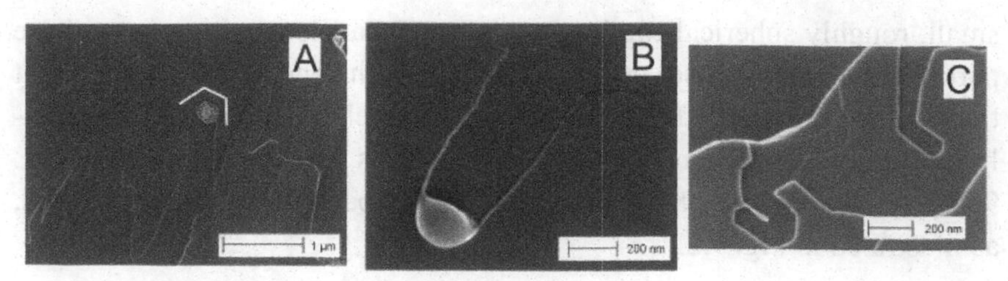

FIGURE 9.24 Small, spherical catalyst particles.

The fairly large particle in Fig. 9.24A cannot be clearly distinguished as having been in the liquid phase during oxidation, yet the tip of the channel is clearly faceted with 120° angles. The particle in Fig. 9.24B was clearly molten during oxidation as it has deposited material on the channel walls. It is interesting to note that since the channel walls have expanded a negligible amount compared to the channel depth, the activity of the catalyst deposited on the wall is significantly less than that of the original particle. Finally a peculiar behavior was found in the partially purified material, where a small catalyst particle is found at the tip of a straight channel, ending in a 120 ° tip (clearly noticeable in Fig. 9.3). The width of the channel is roughly an order of magnitude larger than the particle itself, as seen in Fig. 9.24C. In this case channeling was always found to proceed along preferred crystallographic directions.

Such a wide variety of catalytic behaviors are not unexpected for the natural graphite samples under consideration. Despite being purified, the purification treatments are unlikely to penetrate the graphite particles completely. As such inclusions that may have been trapped within the structure during formation will not be removed and will be subsequently exposed by the oxidation. These impurities can have virtually any composition and hence lead to the diversity of observed behaviors.

In addition to the irregular channels, erratically shaped pits are also found in the natural graphite sample, as shown in Fig. 9.25.

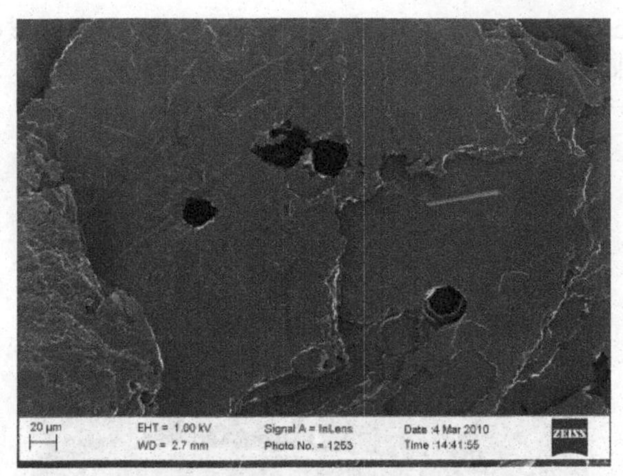

FIGURE 9.25 Pitting in natural graphite (650 × magnification).

Underdeveloped pits are often associated with erratically shaped impurities, as shown in Fig. 9.26 (A and B).

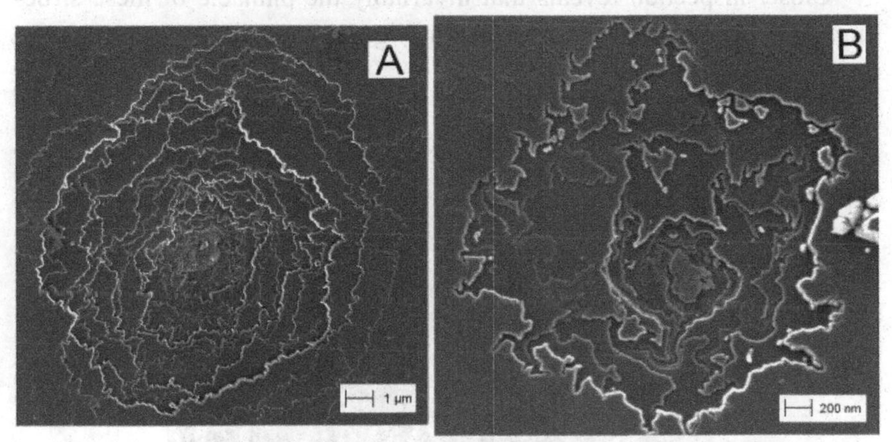

FIGURE 9.26 Impurity particles associated with pitting (35k × magnification).

The myriad of different catalytic behaviors found in this high purity natural graphite sample coupled with the enormous impact catalyst activity has on reactivity, demonstrates the danger of simply checking the impurity levels or ash content as a basis for reactivity comparison. A final morphological characteristic of this material is the presence of spiked or saw-tooth like edge formations, as can be seen in Fig. 9.27.

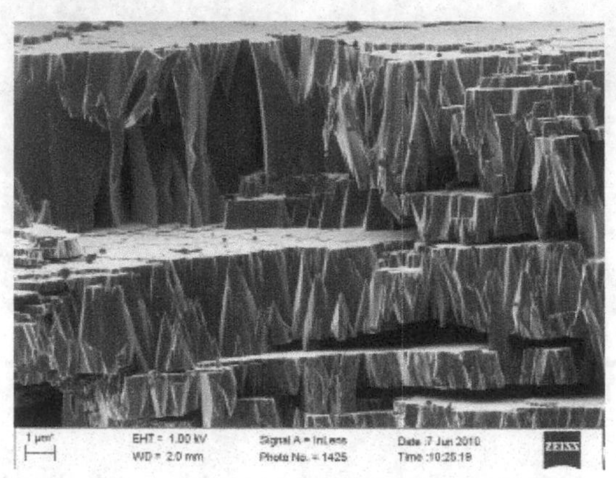

FIGURE. 9.27 Saw-tooth edge formations (15k × magnification).

Closer inspection reveals that invariably the pinnacle of these structures is capped by a particle, as seen in Fig. 9.28.

FIGURE 9.28 Close-up of saw-tooth structures (50k × magnification).

Thus these formations are caused by inactive particles, which shield the underlying graphite from attack. These layers protect subsequent layers leading to the formation of pyramid like structures crowned with a single particle. In some cases as on the left hand side of Fig. 9.29, these start off as individual structures, but then as oxidation proceeds around them, the particles are progressively forced closer together to form inhibition ridges, as can be observed on the right hand side of Fig. 9.29.

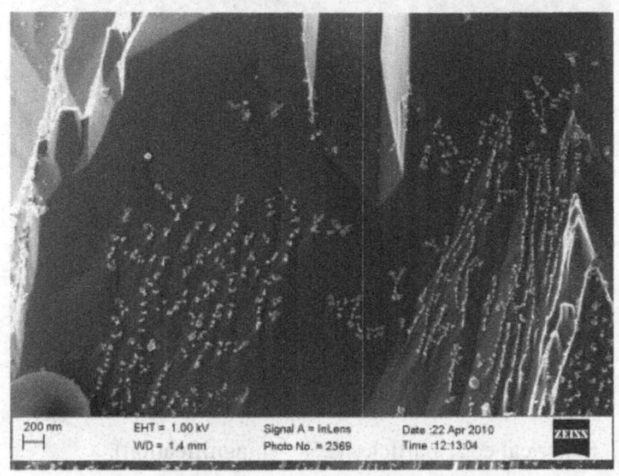

FIGURE. 9.29 Inhibiting particles stacked along ridges (50k × magnification).

In extreme cases these particles may remain atop a structure until it is virtually completely reacted away, for example resulting in the nano-pyramid shown in Fig. 9.30.

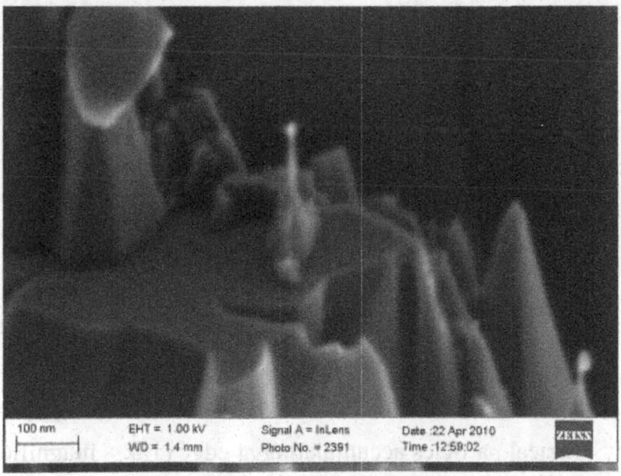

FIGURE 9.30 Nano-pyramid (300k × magnification).

In some cases particles are found which appear to neither catalyze nor inhibit the reaction, such as the spherical particles seen in Fig. 9.31.

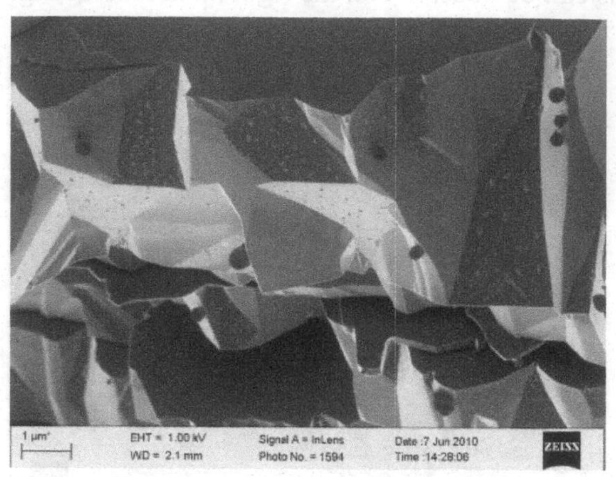

FIGURE 9.31 Spherical edge particles (25k × magnification).

These may be catalyst particles, which agglomerate and deactivate due to their size. The graphite is oxidized away around them, until they are left at an edge, as seen in Fig. 9.32.

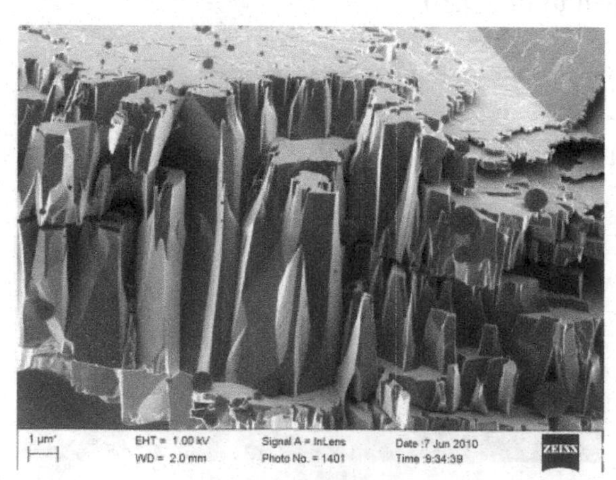

FIGURE 9.32 Spherical particles accumulating at edge (15k × magnification).

The accumulation of inhibiting particles at the graphite edge will inevitably lead to a reduction in oxidation rate as the area covered by these particles begins to constitutes a significant proportion of the total surface area. This may appreciably affect the shape of the observed conversion function, especially at high conversions.

9.3.5 NUCLEAR GRADE NATURAL GRAPHITE

The as-received nuclear grade natural graphite (NNG) exhibits a different morphology from that found in the commercial flake natural graphite. In this case the particles appear rounded and almost spherical, as shown in Fig. 9.33.

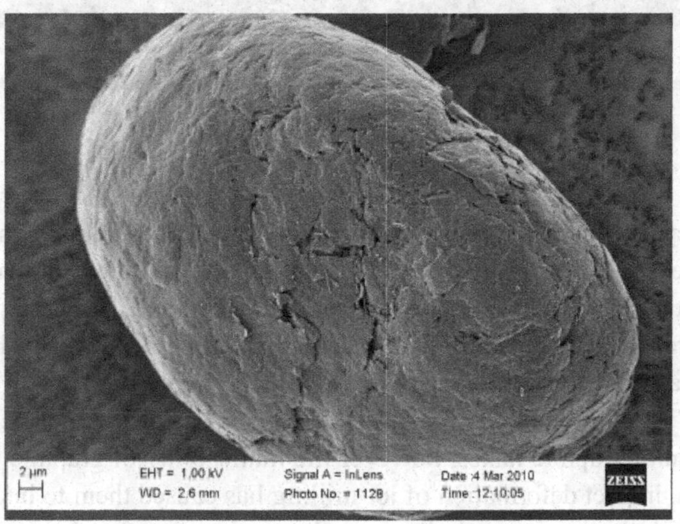

FIGURE 9.33 Rounded nuclear graphite particle (5k × magnification).

When the oxidized NNG microstructures are examined in Fig. 9.34, fairly complex and irregular structures are found.

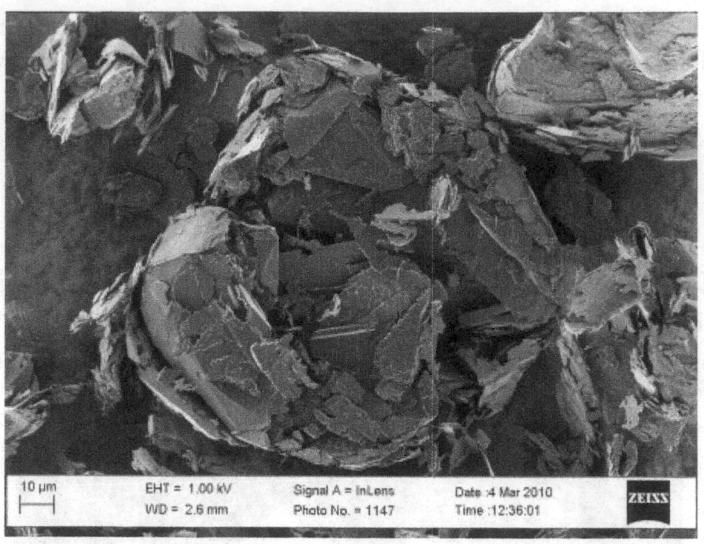

FIGURE 9.34 Oxidized NNG (1k × magnification).

The particles are extensively damaged and crumpled, however the fact that they remain in-tact indicates that this is one continuous fragment. As the outer roughness is removed by oxidation, the multifaceted features of the particle interior are revealed. It may be concluded that these particles are in fact an extreme case of the damaged structure shown in Fig. 9.6. This material has been extensively jet-milled to create so-called "potato-shaped" graphite. Initially the particles may have resembled the commercial natural graphite flakes, however the malleability of graphite coupled with the impact deformation of jet-milling has caused them to buckle and collapse into a structure similar to a sheet of paper crumpled into a ball. Despite the high levels of purification, this material still exhibits extensive catalytic activity, similar to the flake natural graphite, as shown in Fig. 9.35.

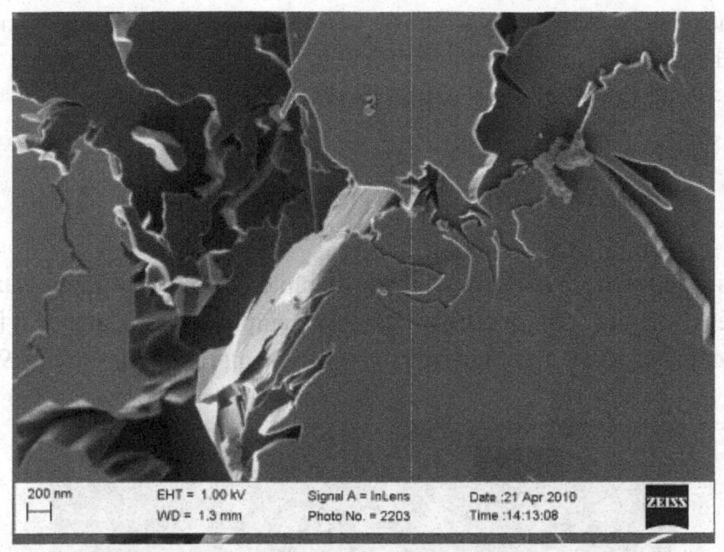

FIGURE 9.35 NNG catalytic activity (50k × magnification).

In spite of the catalytic activity and structural damage, in some regions the basal surface is still fairly smooth and flat across several micrometer, as can be seen in Fig. 9.36, indicating that the material still has good underlying crystallinity.

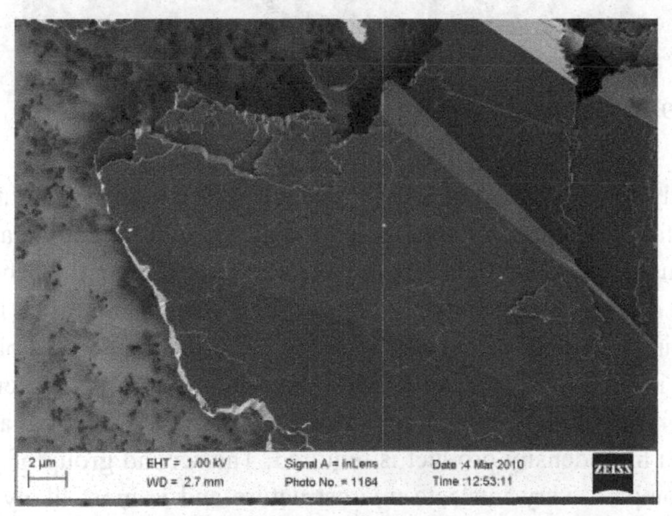

FIGURE 9.36 NNG basal plane (9k × magnification).

Thus this despite being naturally derived and evidently highly crystalline, the microstructure of the NNG material is very complex due to the extensive particle deformation during processing.

9.3.6 NUCLEAR GRADE SYNTHETIC GRAPHITE

The as-received nuclear grade synthetic graphite (NSG) exhibits a remarkably different behavior from the natural graphite samples. At first glance it is possible to distinguish between two distinct particle morphologies in Fig. 9.37.

FIGURE 9.37 Oxidized NSG (700 × magnification).

Firstly, long, thin particles are noticeable with a high aspect ratio. During the fabrication of synthetic graphite a filler material known as needle coke is used. These particles are most likely derived from the needle coke with its characteristic elongated, needle-like shape. This filler is mixed with a binder, which can be either coal tar, or petroleum derived pitch. The pitch is in a molten state when added and the mixture is then either extruded or moulded. The resulting artifact can then be re-impregnated with pitch if a high-density product is required. The second group of particles has a complex, very intricate microstructure and is most likely derived from this molten pitch. They are highly disordered with a characteristic mosaic texture probably derived from the flow phenomena during impregnation. When examined edge-on, the layered structure of the needle coke derived particles is still readily evident, as seen in Fig. 9.38.

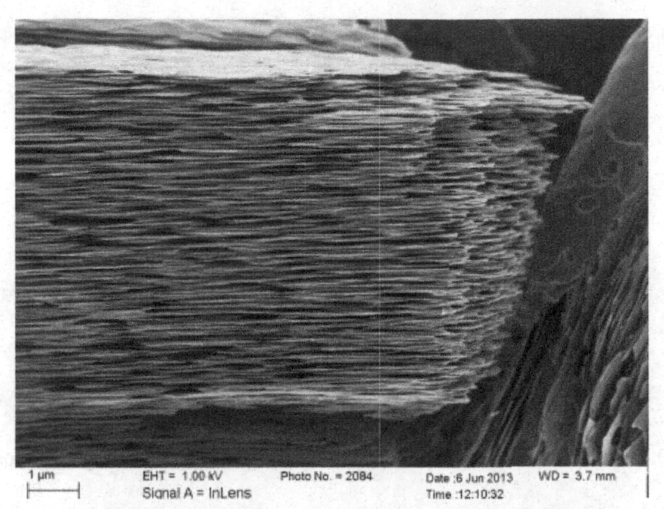

FIGURE 9.38 Oxidized NSG needle particle (20k × magnification).

The needle particles bear some resemblance to the natural graphite flakes, with the basal plane still readily identifiable in Fig. 9.39.

FIGURE 9.39 Oxidized NSG needle particle (4k × magnification).

However, when the basal plane is examined more closely in Fig. 9.40 there is a stark contrast with the natural graphite basal plane. The basal surface is severely degraded, with attack possible virtually anywhere.

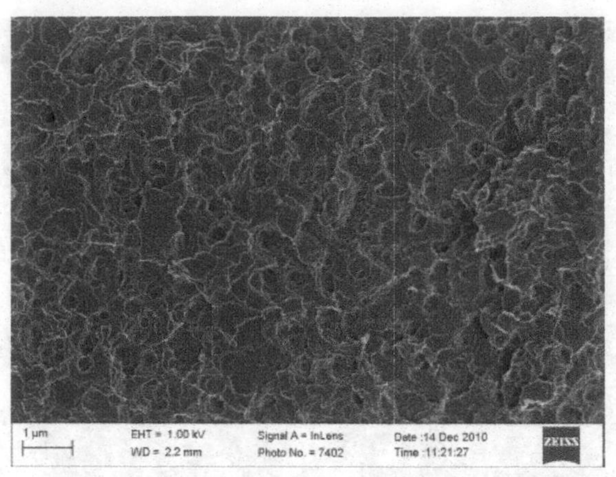

FIGURE 9.40 Oxidized NSG needle particle basal plane (25k × magnification).

The cavities were extensively investigated and no traces of impurities were found to be present. Instead the oxidation hollow has the characteristic corkscrew like shape of a screw dislocation as can be seen from Fig. 9.41. In addition, the pits tend to have a vaguely hexagonal shape.

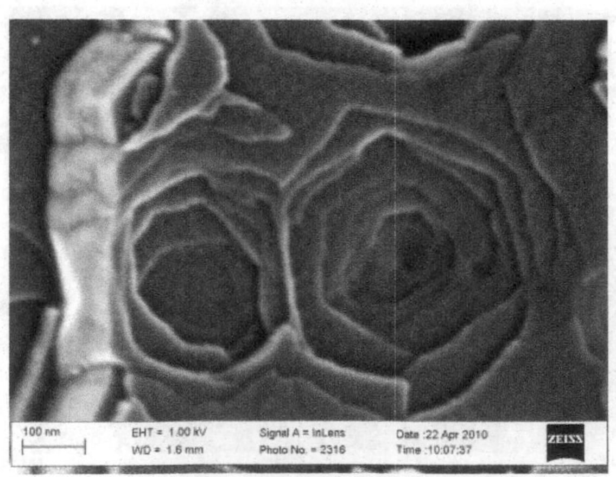

FIGURE 9.41 NSG screw dislocation (320k × magnification).

It is also important to notice that in some regions the defect density is not as high as in others, as can be seen for the different horizontal bands in Fig. 9.42A and also the different regions visible in Fig. 9.42B. This may imply different levels of crystalline perfection in these regions.

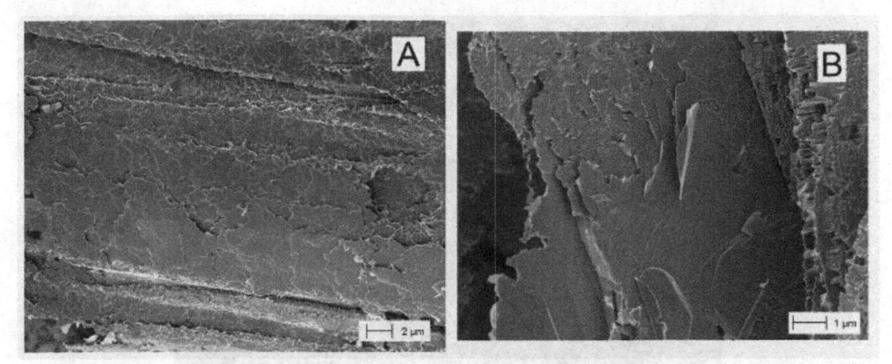

FIGURE 9.42 NSG crystallinity differences (16k × magnification).

When examined edge on as in Fig. 9.43, it can be seen that the needle particles retained their original structure, however any gaps or fissures in the folds have grown in size. This implies the development of complex slit-like porosity, probably initiated by "Mrozowski" cracks, which would not have occurred to the same extent if direct basal attack was not possible to a large degree.

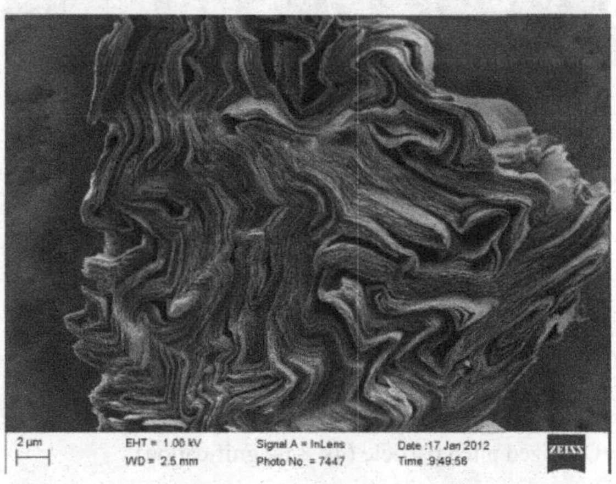

FIGURE 9.43 Slit-like pore development in NSG (8k × magnification).

When the particles edges are examined more closely in Fig. 9.44, the low level of crystalline perfection is further evident. The maximum, continuous edge widths are no more than a few hundred nanometers, far less than the several micron observable in the natural samples, such as Fig. 9.7.

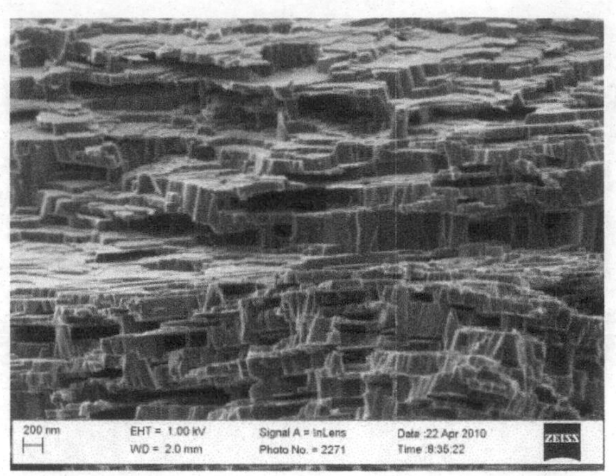

FIGURE 9.44 Degraded edge structure of NSG (50k × magnification).

The complex microstructural development characteristic of this sample is even more pronounced in the pitch particles, as can be seen from Fig. 9.45.

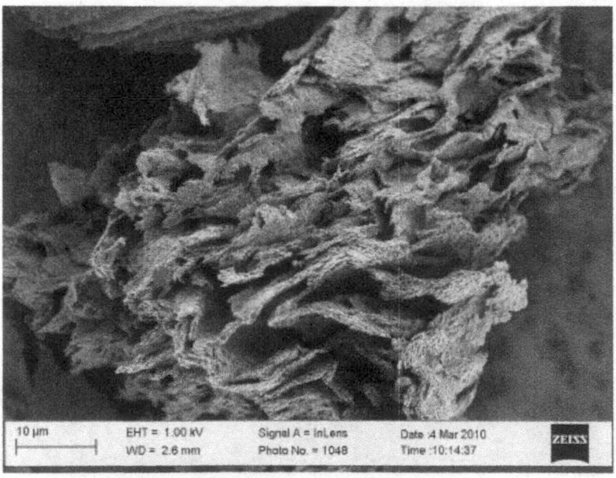

FIGURE 9.45 Oxidized pitch particle (4k × magnification).

These particles lack any long range order, however when their limited basal-like surfaces are examined more closely as in Fig. 9.46, a texture very similar to the basal plane of needle particles is found, indicating possibly similar levels of crystalline perfection.

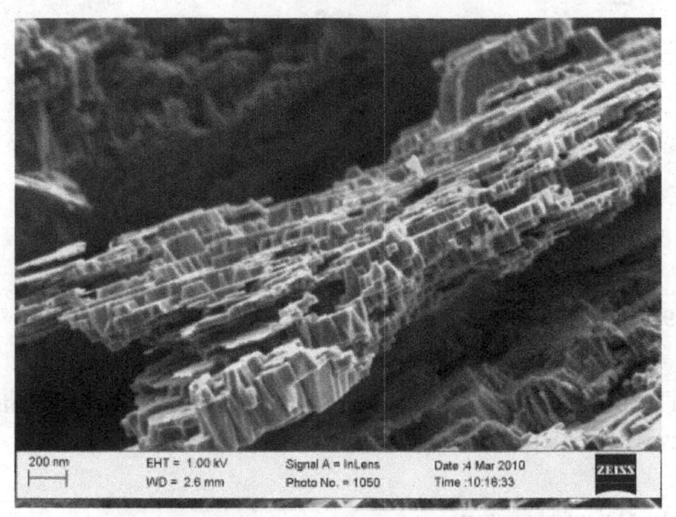

FIGURE 9.46 Oxidized pitch particle surface (88k × magnification).

On the whole, the synthetic material has the most intricate microstructural arrangement and despite the layered nature of the needle coke derived particles being readily evident, the basal surface is severely degraded indicating a high defect density. Thus this graphite can be expected to have the highest inherent ASA of all the samples considered.

9.3.7 REACTIVITY

As a comparative indication of reactivity the samples were subjected to oxidation in pure oxygen under a temperature program of 4 °C/min in the TGA. The measured reaction rate as a function temperature is shown in Fig. 9.47.

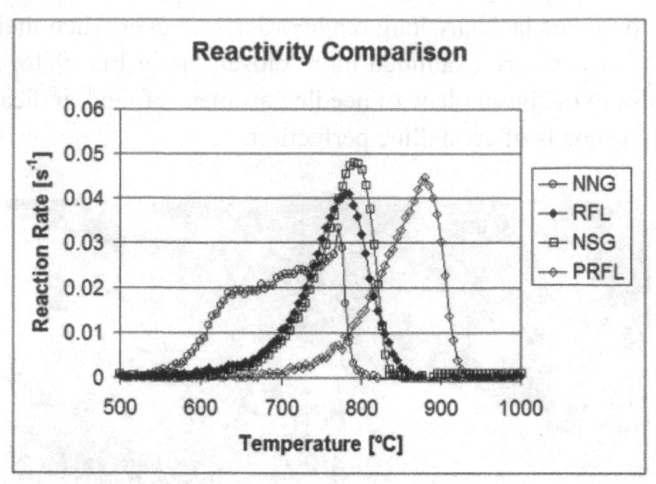

FIGURE 9.47 Reactivity comparison.

As a semiquantitative indication of relative reactivity the onset temperatures were calculated and are shown in Table 9.1.

TABLE 9.1 Onset temperatures.

	Temp (°C)
NNG	572
RFL	696
NSG	704
PRFL	760

It is clear from Table 9.1 and Fig. 9.46 that the NNG sample has the highest reactivity and PRFL the lowest. The NSG and RFL samples have similar intermediate reactivity, although the NSG sample does exhibit a higher peak reactivity. Given the microstructure and impurities found in the respective samples this result is not unexpected. The NNG and NSG samples have comparably complex microstructures, which would both have relatively high surface areas. Despite the higher crystalline perfection of the NNG sample the presence of impurities increases its reactivity significantly above that of the NSG. The sample with the lowest reactivity is the purified PRFL sample, which is not surprising since it exhibited no catalytic activity coupled with a highly crystalline structure and a flake geometry with a large aspect ratio.

The RFL material also has excellent crystallinity and a disc structure with low edge surface area. However, despite its high purity (>99.9%) the RFL sample still contains considerable amounts of catalytically active impurities, thus increasing its reactivity. It is remarkable that despite the comparatively high amount of defects and consequently high ASA, the NSG sample achieves a reactivity comparable to the RFL materials. This indicates the dramatic effect even very low concentrations of catalytically active impurities can have on the oxidation rate of graphite. As can be seen from Fig. 9.2, these minute impurities rapidly create vast amounts of additional surface area through their channeling action. This raises the ASA of the idealized, flat natural flakes to a level comparable to the synthetic material.

9.4 CONCLUSIONS

The active surface area of graphite is important for a wide variety of applications. Through the use of oxidation to expose the underlying microstructure and high-resolution surface imaging it is possible to discern between graphite materials from different origins, irrespective of their treatment histories. This establishes a direct link between the ASA based characteristics, like the oxidative reactivity, of disparate samples and their observed microstructures. This enables like-for-like comparison of materials for selection based on the specific application.

Despite a highly crystalline structure, the oxidative behavior of natural graphite can be dramatically altered through the presence of trace catalytically active impurities and structural damage induced by processing. These differences would be difficult to detect using analytical techniques such as X-ray diffraction, Raman spectroscopy or X-ray fluorescence, due to the similarity of the materials. Synthetic graphite has a much higher defect density than the natural graphite but a similar reactivity to these materials can be achieved if the material is free of catalytic impurities.

In addition, this technique enables insights regarding the extent to which the properties of different materials can be enhanced by further treatments. For example, on the basis of this investigation, it is clear that the oxidative reactivity of the NNG sample may be improved by purification, but due to the damaged structure it cannot achieve the stability observed for the PRFL material, despite both being natural graphite samples. Furthermore, despite having similar reactivities, the NSG and RFL materials have vastly different microstructures and therefore would not be equally suitable for applications where, for example, inherent surface area is very important. In conclusion, given the complexity found in different graphite materials, it is critical that the microstructure should be considered in conjunction with kinetic and other ASA related parameters to afford a comprehensive understanding of the material properties.

This is by no means an exhaustive study of all possible morphologies found in natural and synthetic graphite materials but it does demonstrate some of the intricate structures that are possible.

KEYWORDS

- **catalytically active impurities**
- **crystallinity**
- **microstructural complexity**
- **natural and synthetic graphite particles**
- **particle geometry**
- **processing methods**

REFERENCES

1. Radovich, L. R., Physicochemical properties of carbon materials: a brief overview. In: Serp, P., Figueiredo, J. L., editors. Carbon materials for catalysis, Hoboken, NJ: Wiley; 2009, p. 1–34.
2. Harris PJF. New perspectives on the structure of graphitic carbons. Crit Rev Solid State Mater Sci 2005; 30: 235–53.

3. Luque, F. J., Pasteris, J. D., Wopenka, B., Rodas, M., Barranechea, J. F., Natural fluid-deposited graphite: mineralogical characteristics and mechanisms of formation. American Journal of Science. 1998; 298: 471–98.
4. Pierson, H. O., Handbook of carbon, graphite, diamond and fullerenes. Properties, processing and applications. New Jersey, USA: Noyes Publications; 1993.
5. Reynolds, W. N., Physical Properties of Graphite. Amsterdam: Elsevier; 1968.
6. Laine, N. R., Vastola, F. J., Walker, P. L., Importance of active surface area in the carbon-oxygen reaction. Journal of Physical Chemistry. 1963; 67: 2030–4.
7. Thomas, J. M., Topographical studies of oxidized graphite surfaces: a summary of the present position. Carbon. 1969; 7: 350–64.
8. Bansal, R. C., Vastola, F. J., Walker, P. L., Studies on ultra-clean carbon surfaces III. Kinetics od chemisorption of hydrogen on graphon. Carbon. 1971; 9: 185–92.
9. Radovic, L. R., Walker, P. L., RGJ Importance of carbon active sites in the gasification of coal chars. Fuel. 1983; 62: 849–56.
10. Walker, P. L., RLJ, JMT. An update on the carbon-oxygen reaction. Carbon. 1991; 29: 411–21.
11. Arenillas, A., Rubiera, F., Pevida, C., Ania, C. O., Pis, J. J., Relationship between structure and reactivity of carbonaceous materials. Journal of Thermal Analysis and Calorimetry. 2004; 76: 593–602.
12. Cazaux, J. From the physics of secondary electron emission to image contrasts in scanning electron microscopy. J Electron Microsc (Tokyo). 2012; 61(5), 261–84.
13. Lui, J. The Versatile FEG-SEM: From Ultra-High Resolution To Ultra-High Surface Sensitivity. Microscopy and Microanalysis. 2008; 9: 144–5.
14. Baker RTK. Factors controlling the mode by which a catalyst operates in the graphite-oxygen reaction. Carbon. 1986; 24: 715–7.
15. Yang, R. T., Wong, C. Catalysis of carbon oxidation by transition metal carbides and oxides. Journal of Catalysis. 1984; 85: 154–68.
16. McKee, D. W., Chatterji, D. The catalytic behavior of alkali metal carbonates and oxides in graphite oxidation reactions. Carbon. 1975; 13: 381–90.
17. Fujita, F. E., Izui, K. Observation of lattice defects in graphite by electron microscopy, Part 1. J Phys Soc Japan 1961; 16(2), 214–7.
18. Suarez-Martinez, I., Savini, G., Haffenden, G., Campanera, J. M., Heggie, M. I., Dislocations of Burger's Vector *c*/2 in graphite. Phys Status Solidi C 2007; 4(8), 2958–62.
19. Rakovan, J., Jaszczak, J. A., Multiple length scale growth spirals on metamorphic graphite {001} surfaces studied by atomic force microscopy. American Mineralogist. 2002; 87: 17–24.

[illegible] Derkas, M. [illegible] Natural diamonds [illegible] characteristics and mechanisms of formation. [illegible] 1998, [illegible]
[illegible] Handbook of carbon, graphite, diamond and fullerenes: Properties, processing and applications. New Jersey, USA: Noyes Publications; 1993.
[illegible] Reynolds [illegible] Physical Properties of Graphite. Amsterdam: Elsevier; 1968.
[illegible] Laine [illegible] Walker [illegible] Importance of active surface area in the carbon [illegible] Chemistry. 1963; [illegible]
[illegible] Thomas, J. M. [illegible] graphite [illegible] surfaces [illegible]
[illegible] Carbon. [illegible]
[illegible] journal of [illegible] analysis and [illegible]
[illegible] surface [illegible]
[illegible] graphite [illegible] Carbon. [illegible]
[illegible] Wang [illegible] and [illegible]
[illegible] Harcourt [illegible]
[illegible] of [illegible] graphite [illegible]
[illegible] multiple [illegible]
[illegible] stacking [illegible]

INDEX

A

AASHTO standard, 228
Abundant rainfall, 150
Acac ligand, 165, 174, 180, 182
  Nickel complex, 165, 174, 176
Acetone dioxygenase, 174, 175
Acetophenone, 164, 167
Acireductone dioxygenase, 174, 175
Active surface area, 247, 277
Actual fish catch, 140
Additional models, 14
  equation of state, 14
Advantages of IMAC, 112
  high protein loading, 112
  ligand stability, 112
  mild elution conditions, 112
Aeolotropic, 48
Aerosol, 30
Aerospace structures, 15
Air drag force, 87
Air thermal conditions, 63
Algebraic equation, 82, 86
Alkali metal stearate, 165
Alkaline impurities, 210
Alkylarens, 164, 173, 185
  Cumene, 164, 174, 185
Allocation of carbon, 207, 208, 210
Allometric equation, 65
Aluminum nanoparticles, 10
Aluminum oxide, 10
Ambient parameter, 23
Amino acid chains, 112
  Cysteine, 112
  Histidine, 112, 125, 128
  Tryptophan, 112, 128,
Ammonium salt, 174, 182
Amorphous glassy phase, 191
Analysis of oxidation, 164
Analytical equations, 14
Anatomists, 14
Anisotropic forces, 49
Annual fish catch data, 142
Antoine Lavoisier, 36
Aquatic food web, 139
Aqueous medium, 180
Arabian Sea, 140, 149, 150
Arbitrary Lagrangian Euler, 86
Arc melting chamber, 10
ARD Dioxygenases, 162
Armor, 201
Aspect ratio, 45, 69, 227, 250, 258, 270, 277
Asphalt mixtures, 226–228, 231, 235, 239–241
Asymptotic regime, 28
Asymptotic scaling, 76
Atomic force microscope, 3
Atomistic approach, 14
  direct Monte Carlo simulation, 14
  molecular dynamics, 5, 6, 14, 54, 55, 93
Augite captures, 196
Avogadro number, 64
Axial direction, 2, 79

B

Backbone atoms, 56
Bacterium, 5
Balance of momentum, 36, 54
Ball milling, 9
Band Sequential Format, 142
Bathochromic shift, 179, 182
Bay of Bengal, 140, 149–151
Bead-spring chain, 56
Bead-spring model, 55, 58
BEM advantages, 88

Bounded boundaries, 88
Data input, 88
Integral equations, 85, 87–89
Benchmark solutions, 95
benzene oxidation, 163
Bhatnagar-Gross-Krook, 92
Binding affinity, 125
Biochemists, 15
Biological productivity, 150
Biological systems, 3, 179
Bio-photonics, 13
Blood vessel tissue, 16
Body system, 4
Boltzmann's constant, 51
Bone tissue, 16
Boundary Element Method, 87, 89
Boundary-value problems, 94
Bovine serum albumin, 112, 114, 116, 125
Brownian dynamics, 54, 55
BSA protein, 113
Building blocks, 7
Bulk counterparts, 7
Bulk graphite material, 246
Bulk material, 4, 10
Bulk properties, 7
Burner flame, 216
Temperature peak, 216
Burst of helium, 13

C

Calcium, 208
Capillary tube, 22, 27, 33
Carbon nanotube, 3, 60
Carbon-carbon bonds, 254
Carbonized mass, 217
Carboxylic groups, 112
Cartesian coordinate system, 34
Cartilage tissue, 16
Casting method, 17
Catalytic system, 162, 165, 174, 185
Catalytically active impurities, 246, 277
Celestial mechanics, 82
Galaxies, 82
Planets, 82
Stars, 82
Cell components, 162
Chromosomes, 162
Microtubules, 162
Mitochondria, 162
Ribosomes, 162
Cell height, 172
Cellophane membranes, 112, 113, 117, 125
Cellophane sheets, 113
Cellulose, 113
Glycerol, 114
Ceramic nanomaterials, 10
Cold plasma, 10
Hot plasma, 10
Ceteris paribus, 196
Chelate acac-ring, 174
Chelate ligand, 174, 175
Chemical reaction, 9, 13, 36
Pyrolytic, 13
Photothermal activation, 13
Chemical vapor reaction, 9
Heating heat pipe gas reaction, 9
Chemical vapor, 9, 13
Chemical vapor condensation, 9
Sputtering method, 9
Chlorophyll *a* concentration, 138, 139, 142, 143
Chloroplast pigments, 139
Chromatographic resin, 112
Chromium phase, 209
Citrate molecules, 17
Classical optics, 53
Clinopyroxene, 202, 208, 209
Diopside-augite composition, 202
Olivine, 202
Traces of chromite impurities, 202
Clinopyroxene, 202, 208, 209
Coagulation chemistry, 26
Coal melted, 208
Coarse aorphous structure, 221
Coarse-grained approach, 54
Coastal upwelling, 150
Arabia, 150
Somalia, 150
Coastal zone color scanner, 139
Coating film, 218
Coating intumescence, 223

Coaxial capillary tubes, 27
Coke structure, 220, 221, 222
Collecting plate, 21, 46
Collisions, 5, 12, 40
  impurities of the solid, 5
  vibrating atoms, 5
Colloid chemical methods, 8
Colloidal dispersion, 9
Colloidal metal particles, 11
Command prompt, 142
Commercial flake, 267
Computational fluid dynamics, 92
Computational model, 26, 80
Computational physical process
  Computer simulation, 53-55, 80
  Problem definition, 80
Cone angle, 163
Cone-jet mode, 29
Conical shape, 19, 20
  Taylor cone, 19–24, 27, 90
Constitutive assumption, 48
  Equation of state, 14, 47, 48
Continuum mechanics, 4, 95
Continuum physics, 95
Control-volume analysis, 90
Copper ions, 129
Copper sulfate solution, 115
Core shell nozzle, 84
  Concentric annular pipe, 84
  Cylindrical pipe, 84
Corkscrew shape, 254
Correlation coefficient, 151
Coulomb force, 43, 87
Coulomb's law, 43, 44, 91, 97
Criegee rearrangement, 174
Critical length, 5
  scattering length, 5
  thermal diffusion length, 5
Critical voltage, 23
Crystalline fracture, 203, 206
  Albite, 203
  Feldspar, 203
  Magnesite, 203
  Wollastonite, 203
Crystalline phases, 196
Curvature radius, 163
Cylindrical coordinate system, 34
Cylindrical punching, 202
Cysteine dioxygenase, 175

D

Dangled segments, 62
Darrell reneker, 16
  Dashpot element, 34
Data on structures, 169
Data storage materials, 4
Density functional theory, 6
Design multifunctional materials, 2
Diameter of pipette orifice, 25
  Clogging effect, 25
Diamond diagnostics, 114, 115
  Glucose kit, 114, 116
  Total protein kit, 115, 116
Dichloromethane, 27
Dielectric effect, 23, 24
  Dielectric property, 24
Diffractometer, 202
Digital computers, 86
Dimensionless quantities, 67
Dioxetane fragment, 175
Dipolar interactions, 162
Direct fuel inputs, 138
Distilled water, 114, 116, 216
DNA replication, 17
Drug delivery systems, 16
Dumbbell models, 55
DuNouy tensiometer, 220

E

Earnshaw's theorem, 43
Ekman transport, 150
Elastic dumbbell kinetic theory, 49
Electric heating evaporation method, 9
Electric potential energy, 42
Electrical conductivity, 5, 90
Electrical forces, 22, 41
Electro spun fibers, 21
Electrode material for batteries, 8
Electron donating capacity, 125
Electron-donating ligand, 174
Electrons travel, 5

Electrophiles, 176
Electrospinning dilation, 37
Electrospinning patent, 20
Electrospinning procedure, 2
Electrospinning process
  Charging of the polymer fluid, 20
  Formation of the cone jet, 20
  Instability of the jet, 21
Electrospun nanofibers, 64
Electrospun polymer nanofibers, 75
Electrostatic field, 20, 32
Electrostatic pulling force, 78
Element's angels, 41
El-Nasr Pharmaceutical
  Bovine Serum Albumin, 112, 114, 125
  Copper sulfate, 114, 115
  Ethyl alcohol absolute, 114
  Hydrochloric acid, 114
  Methyl alcohol, 114
  Sodium acetate trihydrate, 114
  Sodium sulphite anhydrous, 114
Elongational flows, 49
Elution conditions, 112, 131
Energy dispersive analysis, 117
Environmental satellite, 139
Epoxy groups, 115, 123
Equator-ward flow, 150
Equivalent mass, 46
Estuarine waters, 139
Ethyl benzene, 162–168, 173, 174, 176, 185
Evacuated chamber, 10
Evacuated quartz tube, 11
Exclusive Economic Zone, 140
Exploding wire method, 9
Extended finite element, 86
Extracellular matrix
  connective tissue, 14

F

Fabricate materials, 96
Fabricating electronic, 7
Fabrication of nanostructures, 9
  bottom–up, 4, 5, 9
  top–down approach, 9
Fabrication technology, 15
Fatigue cracking, 226
Fe-acetyl, 175
FEM Solutions, 88
  discretization, 83, 86, 93
  displacements, 86
  strains and stresses, 86
Feng solution, 61
Ferro fluids, 54, 56
Fiber diameter, 16, 24, 33, 63, 64
  transient regime, 63
  free molecule regime, 64
Fiberglass plastics, 214, 215, 219
Fibrous network, 61
Field of investigation, 5
Film-forming polymer, 222
Filter material, 20
Finite element analysis, 86
  nodal point spatial locations, 86
  restraints, 86
Finite-element code, 91
Fire retardant coatings, 214, 215, 216, 222
Fishing activity, 158
Flake body, 257
Flake geometry, 277
Flat basal surfaces, 250
Flow analysis, 90
Flow of electric charge, 38
Flow of inert gas, 27
  inner tube, 27
Flow path modeling, 46
Flowing gas jacket, 27
Fluid continua, 85
Fluid droplets, 20
Fluid dynamics, 33, 36, 57, 80, 92, 95
Force imbalances, 29
Fourier laws, 36
Fredholm integral equations, 89
Freeze-drying, 18
Fresh water, 149

G

Gal actosidase, 113
Gal adsorption, 125, 127
Gal enzyme, 112, 116, 134

Gas jacket, 27
Gas molecules, 10, 12, 64
ionized, 10, 13
Gas phase, 9, 12
electron beam heating, 9
gas-phase evaporation method, 9
laser heating, 9
plasma heating, 9
Gelation temperature, 18
Geological processes, 258
Geophysical parameters, 139
Geospatial imaging, 142
Ghaffarpour, 227
Giesekus equation, 49, 97
Glass bubbling-type reactor, 164
Glass-crystalline materials, 201
Global oil consumption, 138
Globular molecules, 15
Glycidyl methacrylate, 112, 114
Goddard Space Flight Center, 141
Goniometric method, 220
Grafted membranes, 112, 115, 117, 123, 124, 125
Graphite defect structures, 254
basal dislocations, 254
non-basal edge dislocations, 254
Prismatic edge dislocations, 254, 257
Prismatic screw dislocations, 254
Graphite materials, 246, 247, 248, 277, 278
Graphite sites, 222
Grounded plate, 21
metal screen, 21
rotating mandrel, 21

H

Harmful smells, 22
Harmonic potential, 55
Hartree-Fock methods, 6
Harvesting technologies, 138
HCL solution, 118
Heated substrate surface, 11
Helium gas, 10
Heteroatoms promote, 222
Heterobimetallic, 168, 173, 184, 185
Heteropolar bonds, 209
Heuristic method, 95
Hexagonal shape, 272
Hexamethylphosphorotriamide, 165
Hierarchial data format, 141
High concentration of Silica, 191
High degrees of conversion, 173
High rate of production, 13
High specific surface area, 15
High voltage acceleration, 24
High-speed penetration, 202
Histidine tags, 112
Hollow glass micropipettes, 17
Homogeneous nucleation, 196
Homogeneous state, 191
Homopolar, 209
Hooke's law, 48, 50
Hookean springs, 56
Horizontal bands, 273
Hot mix asphalt, 226, 227, 241
Hydro peroxides, 164, 174
Hydrogen bonding, 162
Hydrophobic core, 19
Alkyl, 19, 183
Hydrophilic, 19, 123
Hydrophobic PGMA, 123

I

IDA ligand, 112
IMAC membranes, 134
Imidazole groups, 125
Immiscible liquids, 190
Immobilized metal ions, 112, 118, 125, 128
Impact breeds, 200
Impact of meteorites, 200
Indian Coast, 140
Indian Ocean, 139, 140
Indigestible plant matter, 15
Inert gas atmosphere, 10
Infinite-time horizon, 92
Influence of viscoelastic, 28
Insofar as electrospinning process, 35
jet, 35
Inter-molecular forces, 18

Internal degrees of freedom, 54, 56, 57
Intestinal peristalsis, 15
Intumescent coatings, 214, 219
Intumescent layer, 222
Iodometry, 164
    Acetophenone, 164, 167
    Methylphenylcarbinol, 164, 167
Iron complexes, 175, 179, 181, 185
Isaac newton's law of gravity, 43

J

Jet diameter, 29, 38, 73
Jet of fluid, 21
Jet thinning, 28, 79, 97
Jet-milled, 268

K

Kinematic viscosity, 46
Kinematical fields, 95
Kutta–Merson method, 84, 85

L

Lactose solution, 116
Lagrange multipliers, 95
Lagrangian axial, 59
Langevin dynamics, 5
Laser chemical vapor, 13
Laser pyrolysis technique, 12
Laser-irradiated material, 14
    boiling, 14
    heating, 9, 14, 208, 209
    melting, 10, 14, 193, 209
    evaporation, 9, 10, 14, 19, 22
Lateral advection, 150
Lattice Boltzmann methods, 92
Lens detection system, 248
Life-stages of fish, 139
Ligament tissue, 16
Light emitting diodes, 6
Light inhibition, 149
    cloud cover, 149
    turbidity, 138, 149
Linear momentum, 39, 64, 91
    vector quantity, 38
Linear morphology, 253
Linear relations, 50
Liquation differentiation, 190
Liquid continuum, 34
Liquid jet, 2, 34, 35, 58
Liquid phase method, 9
    hydrolysis, 9
    spray, 9
Lithium ion batteries, 246
Lithium oxide, 11
Log-conformation techniques, 86
Lorentz force law, 53
Low biological production, 149
Low degree of agglomeration, 12
Low nutrient content, 149
    Low nitrate, 149
    Silicate content, 149
Low-storage methods, 83
Luminescence, 12
Lunar rocks, 190, 192

M

Macro flake structure, 256
Macrocyclic polyether, 180
Macroscopic approach, 14
    hydrodynamic model, 14, 70
Macroscopic-scale structures, 75
Mafic, 201
Magnetic fields, 53, 54, 56
Magnification, 203, 205
    relief, 205
    slip bands, 205
Man-made mineral formations, 201
Marine fish, 139, 142, 158
Mass of beads, 56
Mass spectrometer, 13
Mathematical model, 2, 20, 70, 80, 90
Matlab software, 226, 227, 239
Maxwell's equations, 53, 96
Mechanisms of catalysis, 162
    Binary, 162, 163, 165, 168
    Triple, 162, 184, 185
Melt electrospinning, 32
Melt siminals, 191, 192, 197
Merged spheres, 177
Mesoscopic transport, 5

Metabolites, 175
 Eukaryotes, 175
 Prokaryotes, 175
Metal capillaries, 20
Metal cations, 11
Metal chelate chromatography, 112
Metal organic compounds, 11
Metal rod, 13
Metal vapor nucleates, 10
Metalloligand, 165–167, 173, 185
Metalloproteins, 167, 175
 Copper zinc super oxide, 167
Meteorite impact, 200, 210
Methionine recycle, 175
Methionine salvage pathway, 175
Methyl groups, 125
Methylphenylcarbinol, 164, 167
Metylpirrolidon, 162, 165
Micro fiber cavity, 181, 183
Micromechanical elements, 54
 Beads, 24, 54–56, 59, 113
 Rods, 54, 55, 113
 Platelet, 18, 54
Microporous affinity films, 113
Microprobe analysis, 196, 202, 206
Micro structural, 2, 54, 246, 274, 275
Microstructure of rocks, 190
Millimeter diameter, 31, 33
Mineral aggregates, 190
Mineral alloys, 190, 201
Mixed oxides, 13
 Carbides, 13
 Nitrides, 13
Moderate Resolution Imaging Spectroradiometer, 138, 139
Modern science, 3
Molecular affinity, 4
Molecular anions, 11
Molecular biology, 3
Molecular oxygen, 164, 168
Molten rock, 190
Molybdenum, 11
Monolithic state, 200
Morphologies, 7
Mosaic texture, 271
Moving belt collector, 21
Mrozowski cracks, 273
Multifilament thread, 21
Multifunctional structure, 61
Multi-scale approach, 14
Muscle fiber, 14

N

Nannoplanktons, 15
Nano fiber techniques
 drawing, 16
 phases separation, 16
 template synthesis, 16–18
Nano ZnO for bitumen, 230
 ductility, 227–235, 241
 penetration grade, 227–235, 241
 softening point, 227, 230–235, 241
Nano, 15
 Greek word, 15
Nanoballs, 8
 Dendritic structures, 8
 Nanocoils, 8
 Nanocones, 8
 Nanoflowers, 8
 Nanopillers, 8
Nanocontainers, 7
Nanocrystals, 4, 8
Nanodevice, 3, 4, 5, 7, 8
Nanodisk, 3, 7
Nanoelectronics, 3
Nano-elements, 3, 15
Nanofarad, 15
Nanoplatelet, 3
Nanoporous structure, 4
Nanoreactors, 7
Nanoscale dimensions, 7
 nanobelts, 7
 nanoribbons, 7
 nanorods, 7, 15
 nanotubes, 6, 7, 15, 18
 nanowires, 7, 10, 85
Narrow continental shelf, 149
National science foundation, 15
Natural graphite basal plane, 272
Natural graphite flakes, 246, 250, 268, 271
 Calcareous sediments, 246

Metamorphosed Siliceous, 246
Natural stone, 190
Near-infrared, 139
Needle coke, 270
Coal tar, 270
Filler, 214, 220, 223, 270
Needle-like shape, 270
Petroleum, 270
Neo-Hookean equation, 51
Newton's laws of motion, 39
Newtonian fluids, 28, 49
NNG microstructures, 267
Nonsingular systems, 86
Novelties, 4
Nozzle regime, 28
Numerical models, 27, 54
Numerical relaxation, 77, 92

O

Ocean color data, 138
Oil-water, 190
Optimal graphite amount, 220
Optoelectronic, 7
Orthopedic implants, 16
Ostwald–de waele power law, 49, 97

P

Paper of kowalewski, 89
Parametric analysis, 75
Pelagic species, 138
Penetration of bitumen, 232
Peninsular rivers, 150
Brahmaputra, 150
Ganges, 150
Godavari, 150
Irrawady, 150
Kaveri, 150
Krishna, 150
Mahanadi, 150
Perchlorovinyl resin, 214–216
Perennial river, 150
Peridynamic modeling, 61
Petrurgical materials, 190
Photoinduced process, 13
Phytoplankton, 139
Planar fiber network, 62
Platelet, 18, 54
Poisson probability, 62
Polarized charge density, 43
Polymer edge drops, 17
Polymer solution droplet, 17
liquid fiber, 17
sharp tip, 17
Polymer tensile force, 79
Polysaccharides, 15, 123
cellulose, 15, 113, 114, 125
Porous morphology, 18
Porphyrin linkage, 162
Portable fuel cells, 2
Potential energy, 14, 42
Potential fishing zones, 138
Power-law fluid, 48
Processes of nucleation, 192
Protein helices, 162
Protein molecule blocks, 128
Pulsed laser ablation, 12
Pyrolysis-based method, 12
Pyroxene, 202, 203, 207, 208

Q

Qualitative description, 57
Quantum dot formation, 9
epitaxial growth, 9
Quantum size effect, 7
Quartz, 10, 11, 193, 200, 209
Quartz crystal lattice, 208, 209

R

Radio-frequency, 10
Radius of the jet, 28, 37, 65, 79
Raman spectroscopy, 227
Rapid evaporation, 17, 184
Rapid thermalization, 12
Rapid whipping, 22
Rate of strain tensor, 52
Reagent blank, 116
Redox-active transition-metal, 162
Relaxation method, 77, 91, 92
Resilient modulus, 228
Resonant frequency, 163

RF plasma method, 10
RFL sample, 248, 276, 278
Rheological model, 30
River water freshens, 150
Rock materials, 201
  glasses, 201
  ceramics, 31
Rotating drum collector, 21
Rotational viscometer, 227, 230, 241
Rouse-Zimm chain, 56
Runge–Kutta methods, 83

S

Saline water, 148
Salts QX, 182
Scaffold formation, 16
Scaling laws, 65, 66
Scanning electron microscopy, 196, 202, 247
Science and technology, 3
  biotechnology, 3
  industrial revolution, 3
  information technology, 3
  medicine and healthcare, 3
  national security, 3
Sea surface temperature, 138–141, 150
Semi-inverse method, 94–96
Shell fluids, 85
Skeletal tissue, 16
Slenderbody theory, 76
Smoldering time, 216
Solid phase method, 9
  milling method, 9
  solid-state reaction, 9
  spark discharge, 9
  stripping, 9
  thermal decomposition, 9
Solution droplet, 17
Solution viscosity, 23
Solvent evaporation pyrolysis, 9
  Emulsion, 9
  Oxidation-reduction, 9
  Radiation chemical synthesis, 9
  Room pressure, 9
  Sol-gel processing, 9
Solvent thermal method, 9
South-eastern coast, 143
Spectrophotometer, 115
Spectrum, 2, 5, 14, 139, 179
Spinneret, 21, 22, 32
  Pipette tip, 21
Spintronics, 4
Stishovite peak, 202
Symmetries
  Axisymmetric, 28, 29, 83
  Nonaxisymmetric, 29
Synthesis techniques, 9
  Solid-state process, 9
  Solution precipitation, 9
  Vapor-phase, 9
Synthesize 3D NSMs, 8
Synthetic filament, 14
  Cotton, 14
  Nylon, 15

T

Temperature plasma, 10
Tensile force, 79, 97
Test domain, 26
Textile industry fiber, 14
Thermal expanded graphite, 214, 220, 223
Thermal Lattice Boltzmann Methods, 92
Thermalization, 13
Tissue engineering, 16
Trace channels, 249, 259
Treatment histories, 246, 277
Triple complexes, 166–169, 171, 186
Tropical basin, 140
  North Indian Ocean, 140
Twinning band visible, 255

U

Ultrafine fibers, 26
Ultra-fine nanomaterials, 12
Ultrafine particles, 10
Ultramafic igneous rocks, 201
Ultra-short laser pulses, 14
Upper convected time, 50, 52
Uterine solution, 169, 177, 181, 183
UV radiation, 13

UV-spectrum data, 179

V

Vacuum gauge, 11
Ventilation system, 22
Vertical mixing, 149–151
Vicinity of the air, 42
Virtual test environment, 26
Viscoelastic
  material, 17
  models, 34
  polymers, 79, 97
  stress, 29, 46
Viscometer tests, 241

W

Whipping region, 21, 29
Wind-driven vertical mixing, 150
Wollastonite, 203–205
Worm-like structures, 196
Wound healing, 16

X

X-ray diffraction, 202, 207, 277
X-ray fluorescence, 277

Y

Young-Dupre equation, 220

Z

Zero dimensional nano-element, 3
Zinc oxide, 226, 227

Printed in the United States
by Baker & Taylor Publisher Services

Printed in the United States
by Baker & Taylor Publisher Services